From Clockwork to Crapshoot

a history of physics

From Clockwork to Crapshoot

a history of physics

Roger G. Newton

The Belknap Press of Harvard University Press
Cambridge, Massachusetts • London, England

Printed in the United States of America

First Harvard University Press paperback edition, 2009

Library of Congress Cataloging-in-Publication Data

Newton, Roger G.
From clockwork to crapshoot : a history of physics / Roger G. Newton
p. cm.
Includes bibliographical references.
ISBN 978-0-674-02337-6 (cloth : alk. paper)
ISBN 978-0-674-03487-7 (pbk.)
1. Physics—History. I. Title.

QC7.N398 2007
530.09—dc22 2006043583

To Ruth, Julie, Rachel, Paul, Lily,
Eden, Isabella, Daniel, and Benjamin

Preface

This book is a survey of the history of physics, together with the associated astronomy, mathematics, and chemistry, from the beginnings of science to the present. I pay particular attention to the change from a deterministic view of nature to one dominated by probabilities, from viewing the universe as running like clockwork to seeing it as a crapshoot. Written for the general scientifically interested reader rather than for professional scientists, the book presents, whenever needed, brief explanations of the scientific issues involved, biographical thumbnail sketches of the protagonists, and descriptions of the changing instruments that enabled scientists to discover ever new facts begging to be understood and to test their theories.

As does any history of science, it runs the risk of overemphasizing the role of major innovators while ignoring what Thomas Kuhn called "normal science." To recognize a new experimental or observational fact as a discovery demanding an explanation by a new theory takes a community of knowledgeable and active participants, most of whom remain anonymous. The book is not a detailed history that judges the contributions of every one of the individuals involved in this enterprise, important as some of them may have been, nor does it trace the origin of every new concept to its ultimate source. More modest in scope at the historical micro-level, its focus is on the general development of ideas.

I have greatly benefited from conversations with my colleagues, especially Don Lichtenberg and Edward Grant, and from the resources of the Indiana University library, particularly from the help of Robert Noel, head of the Swain-Hall library. I also thank Holis Johnson, in our astronomy department, for his advice, and my wife, Ruth, for invaluable editorial assistance.

Contents

From Clockwork to Crapshoot

a history of physics

Prologue

For well over six thousand years people have been assembling facts about their world, recording them when possible, and handing them down to their offspring. As this activity evolved into what is now called science, it became more than just a collection of disconnected facts. Important as these remained, they began to serve primarily to anchor a larger conception of the workings and structure of nature, a conception that forms one of the grandest achievements of the human spirit. In the course of this long historical development, as more facts were discovered, concepts changed, sometimes radically. But certain large-scale themes persisted, informing the questions scientists and scientifically inclined philosophers asked and the kinds of answers they expected.

One theme became clearly recognizable early on, and gradually separated the branch now called physics from the other parts of science: it was the ability to predict future events with some confidence of success. Whether dealing with the reliable functioning of instruments of agriculture, engineering, and war or with a description of the course of the awesome heavenly bodies, the goal of understanding was predictability and its causes. The Greek Leucippos articulated this purpose in the fifth century BCE by declaring "Nothing happens without reason, everything has a cause and is the result of necessity." Chinese Taoist writings expressed similar sentiments in the third century BCE: "All phenomena have their causes. If one does not know these causes, although one may be right [about the facts], it is as if one knew nothing." By the time of Aristotle, the explicit goal of physics (or "physiology," as it was then called) was to understand causes and to use this understanding for prediction. Aristotle himself set down laws governing the movements of the heavenly bodies and rules determining the motions of objects on earth.

Surviving the vicissitudes of history, including the fall of the Greek and Roman civilizations and the rise of Christianity and Islam, the effort to understand nature by discovering fundamental laws culminated in the great scientific revolution led by Galileo Galilei and Isaac Newton. Aided by Newton's new mathematics, determinism found its apex of expression at the end of the eighteenth century in Laplace, who foresaw a science to which "nothing would be uncertain and the future, as the past, would be present to its eyes." However, since we could not possibly know enough to make such all-encompassing predictions, we would have to make do with probabilities, and he provided the needed mathematical tools. Before long, the pervading determinism was further eroded: the second half of the nineteenth century witnessed the direct introduction of the alien concept of probability into physics, and probability dominated basic science throughout the twentieth century. At the most fundamental level, chance took the place of necessity.

But another equally important change accompanied the shift from causality to probability, as we shall see. With reliable prediction of future events at the submicroscopic level no longer within reach, the focus of explanatory theories moved away from their previous Aristotelian purpose of postulating laws of motion to a rather more Platonic goal of explaining structure. In a certain sense, *How?* questions were replaced by *Why?* Since the question "How does an electron move?" could not be answered, it was replaced by "Why does an electron exist?" and "Why does it have the mass it has?" Physics began to explain the detailed properties of matter, both in bulk and in the form of its constituents.

The change in the explanatory aims of physics during the last half century goes even further. Physicists no longer consider it sufficient for physical understanding to point to a force with certain characteristics—or in more modern terminology, an interaction—as Newton had done for the solar system with the force of gravity. They believe that a unified and grand "final theory" will follow logically from a general abstract principle of symmetry, which they accept as an axiom, or else will simply arise because it is the only mathematically possible way for nature to be. If Galileo thought that mathematics was the language of nature, some physicists now go even further, believing that the laws of nature should be mathematical theorems. The course of this development will be outlined in this book.

one

Beginnings

When and where what we now call science started, nobody knows. But we do know that it slowly grew from humans' awe at the heavens and from the arts of healing, hunting, construction, and war. Seeds of it appeared during the Stone Age, in the invention of such hunting implements as the bow and arrow. While the early hunters surely did not try to understand the flight of the arrow, the arrow's effectiveness depended on the predictability of its flight, which was taken for granted. Similarly, agriculture owed its development to experimentation and reliance on previously observed outcomes. Very early hints of what eventually grew into medical science appeared in Mesopotamia—roughly the region now making up Iraq and Syria—and in Egypt, where circumcision was practiced as early as 4000 BCE. Signs of its use have been found in bodies exhumed from prehistoric graves, and the operation is clearly depicted on the walls of a tomb of the sixth dynasty in Egypt, which ruled c. 2625–2475 (Fig. 1). Imhotep, the earliest physician known by name—an astronomer and an architect as well, later venerated as a god—lived about 3000 BCE. Good archeological evidence also indicates that trepanation—cutting out disks from the skull—was performed on living people in prehistoric times. And some of them survived the procedure—we know this because living bone tends to heal itself, and new growth has been unambiguously identified on some of these skulls. Why and how this delicate operation was done is unknown.

Interest in the heavens served less practical purposes and may therefore perhaps be regarded as the proper precursor of "pure science," but that interest seems always to have been closely associated with religion. Individual stars were identified either with specific gods or with the homes of deities. Capricious as the gods were in general, the regular movements of their celestial images offered a reassuring sign of order, while unusual events such as eclipses of the moon or the sun were frightening disruptions of that order. Anyone able to predict these unsettling phenomena was regarded as possessing extraordinary powers. Though it could be found in other places as well, an intense interest in the movements of the heavenly bodies is definitely known to have existed quite early in Egypt, as the devel-

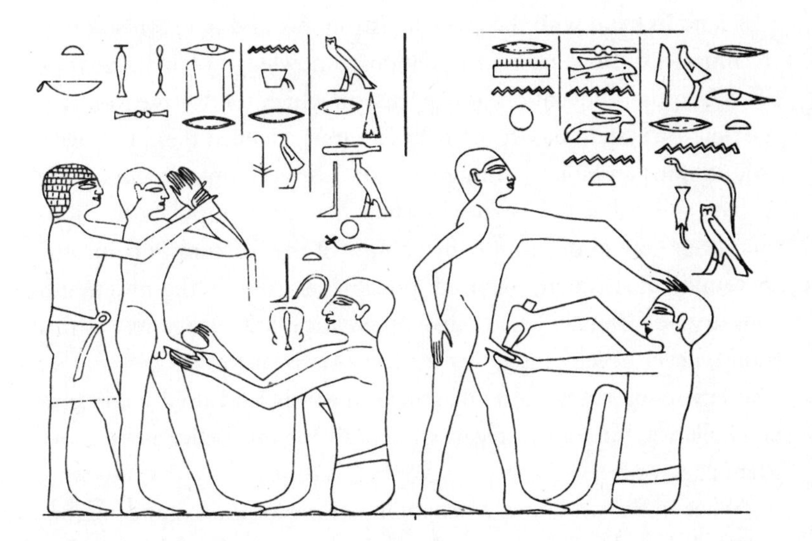

Figure 1 A representation of circumcision with a stone knife, beginning of the Sixth Dynasty in Egypt. (Sarton, *Introduction to the History of Science*, vol. 1, p. 43.)

opment of its stellar-based calendar clearly shows; this calendar can be traced back as far as 4236 BCE. (Note the precision, probably the earliest known date in history.) The construction of a calendar, of course, is the surest sign of faith in the regularity of daily life.

Along with such proto-scientific learning, Egyptian technology became increasingly sophisticated as well. The measurements of blocks used for the pyramids constructed during the thirtieth century BCE are remarkably precise; for example, the leveling of a 50-foot beam was done correctly with an error of only 0.02 inches. "The accuracy of three granite sarcophagi of Senusert II, Twelfth Dynasty (2000–1788), averages 0.004 inches from a straight line in some parts, 0.007 inches in others," according to the science historian George Sarton.[1] Such accuracy implies considerable practical knowledge of stereometry (measurement of solids). Similarly, the construction and erection of enormous Egyptian obelisks required not only practical skill but also great technical know-how.

Hand in hand with the study of astronomy and the acquisition of technical expertise came a developing knowledge of mathematics. The Egyptians' invention of papyrus, on which they wrote in hieroglyphics (a system of writing with pictures), enabled them to leave a voluminous and long-lasting record of their accomplishments. Early papyri show that the Egyptians knew how to manipulate fractions and even how to determine the volume of the frustum (a truncated pyramid) of a square pyramid, possibly as early as the nineteenth century BCE. But they let their mathematical knowledge rest at this point, never developing it any further. (Many historians believe that the Egyptians knew the Pythagorean theorem and used it for land surveillance, but Sarton regards the basis for this belief as no more than guesswork.)

What is known about Mesopotamian proto-science and mathematics is based on modern readings of a large number of clay tablets bearing cuneiform inscriptions in the Sumerian and Accadian languages. Cuneiform employs wedge-shaped signs incised by means of a reed on soft clay, quite different from Egyptian hieroglyphics. The

Sumerian civilization flourished from the beginning of the third until the middle of the second millennium BCE, and its number system was position-based, analogous to our own. In such a system the value of any numeral in a given number depends on its position in relation to the other numerals. But since the Sumerians lacked a zero, their numbers were sometimes ambiguous; and after an early period in which the decimal system was used, they employed a peculiar mixture of bases 10 and 60. The Egyptians' number system, by contrast, though based on 10 like our decimal system, was analogous to Roman numerals, in which position played little or no role; this system was a dead end, because it did not allow arithmetic to develop. The principle of position had to be reintroduced in Europe a thousand years later by way of India, where it had been in use at least since the third century BCE.

The Mesopotamian civilizations directed their quasi-scientific attention primarily toward astronomy and commerce. Sumerian astronomers began by constructing a lunar calendar, which they later modified, assuming the year consisted of 360 days and dividing the day into 12 equal hours, the legacy of which echoed through the ages. But the greatest astronomical achievements of the later Babylonians were many extremely detailed lunar and stellar observations, and in particular an accurate tabulation of the rising and setting of the planet Venus. Living in the fourth century BCE, the astronomer Kidenas (also known as Kidinnu) is believed by some historians to have discovered the precession of the equinoxes—the slow circular motion of the point in the sky above the North Pole (approximately the position of the North Star) about which the whole body of stars in the northern hemisphere is seen to rotate once every 24 hours. There is a similar circle above the South Pole in the southern hemisphere. As a result of this slow motion, the intersection of the plane of the earth's equator with the ecliptic (the plane of the orbit of the earth about the sun) rotates as well, and this intersection determines the equinoxes (thus the name).

As we now know, the motion is caused by a precession of the axis of rotation of the Earth, so that the inclination of this axis with respect to the ecliptic wobbles with a period of about 26,000 years. Historians therefore have good reason to regard the Babylonians as the founders of an early form of scientific astronomy. Two hundred years later, the precession of the equinoxes was clearly discovered by the great Greek astronomer Hipparchus, albeit relying not only on his own observations but also on early Babylonian star data, without which he could not have made the discovery.

The many unearthed tablets recording business transactions, inventories, payrolls, and accounts testify to the strong interest of the Sumerians in matters of trade. As a result of this preoccupation, they made important advances in problems connected with weights and measures, focusing their mathematical attention primarily on arithmetic, at which they excelled. For example, one of the tablets, of c. 2000 BCE, solves the problem of calculating how long it would take for a given sum of money to double at 20 percent compound interest. Others show that they were able to solve not only simultaneous linear equations for many unknowns but two simultaneous quadratic equations for two unknowns, as well as some special cubic equations (though they did not actually use equations as such). They even manipulated negative numbers, a facility that Europe did not acquire until more than three thousand years later.

In geometry, the Babylonians knew the areas of right and isosceles triangles, as well as the volumes of a rectangular parallelepiped (a solid with six faces, each a parallelogram), of a right circular cylinder, and of the frustum of a square pyramid. There is also convincing evidence that they had some knowledge of the Pythagorean theorem, but in circular measurements they were behind the contemporary Egyptians: the Egyptian value of 3.16 for π was closer to its correct value of approximately 3.14 than the Babylonian value of 3.

The manufacture of glass, pottery, and glazes, as well as paints, drugs, cosmetics, and perfumes, was a precursor of the science of chemistry, and the Sumerians produced all of these. Archeologists have even found a remarkable small cuneiform tablet that contains an actual recipe for the creation of a glaze. By far the earliest record of its kind, it dates from the seventeenth century BCE, and archeologists have unearthed nothing like it from the next thousand years.

The famous Code of Hammurabi contains, among its lengthy list of laws and regulations, a specific pay schedule for the performance of various surgical procedures on persons of different ranks. King Hammurabi ruled Babylonia in the first half of the eighteenth century BCE, and his specified fees indicate that surgeons were able to perform their art with bronze lancets on various parts of the body, including the eye. Later Greek sources as well as Egyptian documents going back to the fourth millennium show that the practice of medicine, in both Babylonia and Egypt, was extremely specialized, with different specialists for each part of the body and each disease. Clay models of the liver made by Babylonians and Hittites—a civilization that flourished in the second millennium in Anatolia and subsequently spread to Mesopotamia—can be seen in various museums around the world. Internal medicine in Babylonia was closely allied with people's religious beliefs, and physicians worked hand in hand with priests, relying heavily on incantation and divination.

The gradual replacement of bronze by the much harder metal iron in the Mesopotamian and Egyptian region produced a great upheaval that lasted for several centuries around 1000 BCE. The advantages of the new iron weapons were quickly exploited by their possessors, shifting the centers of power but leaving little time for the disinterested acquisition of knowledge. As a result, further developments that might have led to advances in science or quasi-science were severely disrupted and had to await a rebirth, which eventually took place in the Aegean area of the Mediterranean. No other civilization anywhere else in the world, so far as historians know, developed a comparable level of knowledge at a time prior to 1000 BCE.

During the middle of the second millennium, the eastern Mediterranean had been dominated by the Minoan culture centered on the island of Crete. Whether the demise of this flourishing civilization should be attributed primarily to the gigantic eruption of the volcano Thera on the nearby island of Santorini is still a matter of controversy, but the tardiness of the Minoans in adopting iron technology led to their defeat, first by the Dorians from the north, followed by the Phoenicians from the south, who continued to colonize and dominate the entire Mediterranean coast.

The greatest contribution of the Phoenician civilization, without which the subsequent development of Greek culture and all of Western science surely would have been impossible, was the invention of the alphabet. Even the Hindus learned the art of alphabetical writing from the Phoenicians. In its original form it had no signs for short vowels—and neither Hebrew nor Arabic have such signs to this day—but when the Greeks imitated the Phoenician alphabet, they added symbols for these vowels as well. The newly acquired writing skills enabled them to advance beyond the oral Homeric lore.

As we enter the era of Greek civilization, the state of proto-scientific knowledge in the world may be characterized as descriptive, with aims that were mostly practical or technological, but in part also religious and mystical. In the areas of medicine, agriculture, warfare, hunting, and construction, the accumulated know-how served the purpose of making daily tasks easier and more reliable. Careful observers of the regularity of celestial bodies used their powers of prediction either for reassurance or for further mystification. At this stage, no attempt was yet being made to understand or explain nature.

two

The Greek Miracle

Many of the proto-scientific ideas of the early Greeks had their roots in the Egyptian and Babylonian traditions they inherited. This included the art of astronomical observation and a knowledge of specific regularities such as the (approximately) 18-year cycle, called the *saros,* which brought the moon and the sun in the same relative position, enabling the more or less reliable prediction of eclipses. However, the desire to go beyond observations of regularity and to look for a rational explanation for the movements of heavenly bodies—the beginning of astronomy as a science in the modern sense—seems to have been typically Greek. It differed substantially from the occult and mystical astrology that had come down to them from the Babylonians and Egyptians and which continued to retain its appeal for millennia, even to this day.

Real science was just one component of a sudden emergence of high culture, beginning in the sixth century BCE, that is sometimes hailed as the "Greek miracle." The cradle of this miracle was located in Ionia, on the west coast of Asia Minor (Fig. 2). Ionia was the meeting place of many caravan routes from beyond the Black Sea, from Mesopotamia, and from Egypt, and of sea trade with the Aegean islands and all of the eastern Mediterranean. Miletos, one of the

ITALY
Cumae
Elea
Tarentum
Sybaris
Thurii
Crotone
Locri
Reggio
Messina
Naxos
Etna
Catania
Leontini
Syracuse
IONIAN SEA
Epidamnus
Illyria
Apollonia
Corcyra
Dodona
Macedon
Mount Olympus
2917 m
Thracia
Abdera
Stagira
Chalcidice
Larissa
Thessaly
Iolcos
Pagasae
Phthia
GREECE
Opus
Delphi
Euboea
Chalcis
Thebes
Megara
Athens
Corinth
Elis
Olympia
Phlius
Mantinea
Argos
Aegina
Ceos
Delos
Sparta
AEGEAN SEA
Byzantium
Propontis
Cyzicus
Lampsacus
Troas
Phrygia
Troy
Aeolis
Mysia
Mytilene
Lesbos
Lydia
Phocaea
Smyrna
Sardis
Chios
Clazomenae
Teos
Colophon
Ionia
Ephesus
Samos
Mycale
Priene
Miletus
Caria
Halicarnassus
Cos
Cnidus
Doris
Ialysus
Camirus
Lindus
Rhodes
0
200 km
100 mi

principal Ionian harbors, grew wealthy from the numerous colonies it had set up along that coast, and that city had the additional good fortune of giving birth to Thales, a man remembered long after his death in about 547 BCE. His fame was based in part on an enduring but no doubt apocryphal legend that he had correctly predicted (using the saros) the solar eclipse of May 28, 585, which occurred in the middle of a stand-off between the armies of the Lydians and the Persians. The sudden darkness so impressed the two kings that they ceased fighting and made peace. Whereupon the oracle of Delphi pronounced the person who had had the knowledge and wisdom to predict the event a wise man, and Thales of Miletos remained forever included among the otherwise variable group of legendary Seven Wise Men of the early Greek tradition. To be able to successfully foretell such an important phenomenon as a solar eclipse was a sign of the greatest intellectual power.

Benefiting from extensive travels to Egypt, Thales was both the first Greek mathematician and the first Greek astronomer, but he was also a very practical politician and businessman. Aristotle records the tale that Thales, after predicting that next year's weather would produce a large olive harvest, proved his business acumen by immediately buying up all the olive presses he could lay his hands on, making a fortune when his prediction came true and he could lease them out at great profit—the first scientific entrepreneur on record. As an engineer, Thales is credited by Herodotus with diverting the River Halys to allow King Croesus's army to cross safely, thereby demonstrating that the flow of rivers was not governed by the gods.

On the basis of what he had learned during his travels in Egypt, Thales was able to perform such geometrical tricks as accurately estimating the height of buildings and the distance of ships from shore. What sets him apart is that he did not stop there but tried to un-

---

Figure 2 *(opposite page)* Ancient Greece.

derstand the principles underlying the solutions of such practical problems and to explain them; he was the founder of geometry as a science. Whether he was actually able to prove all the geometrical principles and theorems he discovered we do not really know. Nevertheless, he was the first person anywhere who is known to have understood that such general theorems were needed, and that explaining practical solutions to problems on the basis of general principles is of far greater value than merely solving individual cases. He was therefore the first real scientist, and his influence on later Greek developments was profound.

Just as Thales was not satisfied with solving certain specific geometrical problems without insight into the underlying principles, so he tried to understand the world by asking what it was ultimately made of—and his conclusion was that the basic stuff was water. If that strikes us as naive, it was not really so unreasonable, Sarton points out, considering the life-giving and life-renewing role that water played in the Mediterranean climate.

Fifteen years his junior, Anaximander of Miletos shared Thales's strong desire to understand the nature and workings of things. What we know from his few writings left to us is that in the case of the motion of the sun he pursued this search for understanding by means of detailed observations at various times of the day and seasons of the year, recording the length and direction of the shadow cast by a gnomon, a vertical stick stuck in the ground with a clear horizontal area around it (or an isolated, tall pointed vertical artifact such as an obelisk). This instrument had already been used for marking the time of day by the Babylonians and the Egyptians, but Anaximander employed it for serious astronomical purposes, determining the dates of equinoxes and solstices. In his theories concerning the constitution of the world (including both the nature of the earth and the evolution of life), he shared the desire, which permeated the emerging Greek science, to explain it all by means of a small number of general laws. However, in order to avoid some of the difficulties with

Thales's theory that everything is ultimately made of water, he took refuge in an abstract, metaphysical concept he called *apeiron,* the nature of which remained obscure.

Anaximander's successor in the quest for a primary substance, Anaximenes of Miletos, returned to the more down-to-earth spirit of the Milesian tradition by postulating air, or the wind, as permeating and underlying everything. None of his writing has survived, but his approach seems to have appealed to his fellow Milesians more than Anaximander's.

Looking around the world for contemporaries of these early Ionian scientists (or, as they called themselves, physiologists), we find Confucius in China, who subsequently exerted an enormous, long-lasting influence. However, as Joseph Needham informs us in his magisterial *Science and Civilisation in China,* the contributions of the Confucians to science were almost entirely negative. Chinese civilization had to wait another three centuries for the Taoists, enemies of the Confucians, to equal the Ionian insights into nature. Opposing the kind of scholastic pursuit of knowledge of the Confucians, based as it was on book learning divorced from the world of nature and concerned primarily with matters of rank and privilege among the feudal society of the day, the Taoists pursued the "Tao of Nature" by observation. The legendary Tao philosopher Lieh Tzu, whose very existence is uncertain, is said to have discussed cosmology and the infinity of time and space in his work.

Near the end of the period of the Warring States, during the reign of the first emperor Chhin Shih Huang Ti, scientists under the patronage of Lü Pu-Wei compiled a set of Taoist writings entitled *Lü Shih Chhun Chhiu.* Completed in 239 BCE, this compilation reveals a spirit not unlike that of the early Greek physiologists: "All phenomena have their causes. If one does not know these causes, although one may happen to be right [about the facts], it is as if one knew nothing, and in the end one will be bewildered . . . The fact that water leaves the mountains and runs to the sea is not due to any dislike of the mountains and love for the sea, but is the effect of height as such . . . Therefore the sage does not inquire about endurance or decay, nor about goodness or badness, but about the reasons for them."[1] Needham, "had no doubt that" the main motive of the Taoist philosophers in wishing to engage in the observation of nature was "to gain that peace of mind which comes from having formulated a theory or hypothesis, however provisional, about the terrifying manifestations of the natural world surrounding and penetrating the frail structure of human society . . . This distinctly proto-scientific peace of mind the Chinese knew as *ching hsin.* The atomistic followers of Democritus and the Epicureans knew it as . . . ataraxy."[2]

In the contemporary Hindu realm, Prince Gautama, the Buddha, who lived from c. 560 to c. 480, developed a philosophy based on agnosticism, rejection of superstition, and respect for reason and truth. His teachings, which developed into a widespread religion, would have made an excellent basis for science, had it not been for the Buddha's total lack of scientific curiosity about the world. Jainism, another heretical offshoot of Brahmanism started during the Buddha's lifetime, found physical science more congenial and postulated that matter was composed of atoms with a variety of qualities. This qualitative doctrine had little in common with the atomism developed by the Greeks.

Let us then return to Greece and the sixth century, but this time to the western part of Greece, which saw a great flowering of religion hand in hand with the emergence of science. The man who embodied this fertile combination was Pythagoras, about whose life there exist only dubious accounts written long after his death. Born on Samos, an island near Miletos, he may have been a student of Thales, who supposedly recognized his genius at an early age. As an adult, Pythagoras is said to have fled the tyranny of Polycrates, the ruler of Samos, traveled much in Babylonia and Egypt, and eventually settled in the Dorian colony of Croton, near the southern tip of Italy. Surrounding himself there with a community of male and female disciples who shared his secrets, observed his dietary taboos, and lived simply and poorly, Pythagoras established an influential cult of religious mystics. The secrecy of the Pythagorean religion prevented many of its details, as well as those in the life of Pythagoras himself, from being known. After his death in about 500 BCE his cult continued to exist for another fifty years before it was suppressed.

Following the Milesian tradition of Thales—namely, establishing a theorem rather than remaining satisfied with knowing special instances—Pythagoras is, of course, generally best known for his famous theorem in geometry: the square of the hypotenuse of a right triangle is equal to the sum of the squares of the other two sides. The proof he is said to have given is very simple and is still taught in high schools today (Fig. 3).

Pythagoras also established a number of other geometrical theorems and contributed extensively—largely by speculation—to arithmetic. For example, he appears to have been the first to make a distinction between even and odd numbers. This distinction seems trivial to us now because in our number system even and odd numbers are easily recognized by their last digits. At the time of Pythagoras, however, the use of literal numerals by the Greeks was still very cumbersome. His primary arithmetical tools were drawing dots in sand, and counting out pebbles. (The word *calculate* stems from the Greek word for pebble.)

The idea of a spherical earth probably originated with Pythagoras, and it seems to have been based primarily on the observation that distant ships become gradually visible on the horizon from the tops of their masts down. However, at the center of Pythagorean philosophy was the concept of spheres and circles as ideal shapes, and that surely must have influenced him as well, just as Plato's analogous views exerted their influence later on the astronomical ideas of Ptolemy. Pythagoras explained the movements of the planets in terms of uniform motions on individual circular orbits. No longer satisfied with merely describing these motions by means of numerical tables,

Pythagoras wanted to understand them, a seminal scientific step. At the same time, the notion that heavenly events were governed by orderly laws and by the gods, while those on earth were subject to chaos, disorder, disease, and death, engendered a philosophical dualism that cast a shadow over both science and religion for millennia.

Another part of science to which Pythagoras significantly contributed would turn out to be an early precursor of acoustics. System-

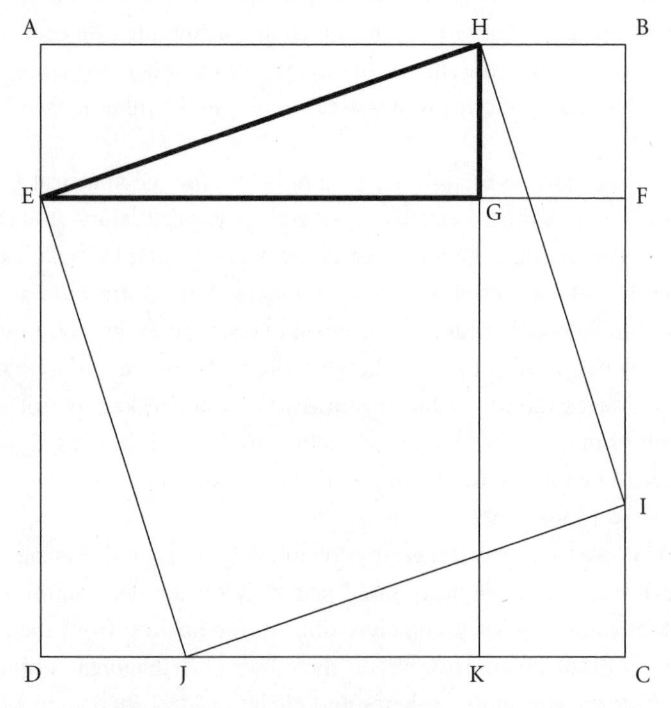

Figure 3 Proof of the theorem of Pythagoras. For a given right triangle EGH, construct the square ABCD whose side-length equals the sum of the lengths EG + GH. Then, (ABCD) = (EGKD) + (HBFG) + 2(AHGE) = (EHIJ) + 4(AHE), if (ABCD) denotes the area of the figure enclosed by ABCD. But (AHGE) = 2(AHE); therefore 2(AHGE) cancels 4(AHE), and what remains is (EHIJ) = (EGKD) + (HBFG).

atically experimenting with string instruments, such as the lyre and the cithara, which had been long known in Babylonia and Egypt as well as in Greece, he discovered that the sounds produced by strings whose lengths were in ratios of 12:9:8:6 were harmonious. In his customary generalizing manner he concluded that numbers were the essence of all things, and proportions in particular played a mystical role. Transferring these notions to the celestial sphere, subsequent astronomers hailed Pythagoras for making "the universe sing in harmony." Both the important later role of mathematics in science and the abuse of mystical numerology by pseudo-scientists and quacks can be traced back at least in part to the influence of the Pythagoreans.

As we enter the fifth century BCE, the first major Greek scientist we encounter is Empedocles (c. 492–c. 430), born in the Sicilian city of Agrigentum, which was then a beautiful and wealthy center of culture before being destroyed by the Carthaginians in 406 BCE. In addition to making a number of contributions to medicine, such as identifying the labyrinth of the ear, recognizing the importance of the skin as well as the heart for respiration, and appreciating the role of the blood vessels as the bearers of heat, Empedocles played a social role in what we would now call public health, and he also performed physical experiments and made systematic observations. On the basis of these activities, he came to the conclusion, which remained extremely influential in one form or another for a very long time, that everything is made up of four elements—fire, air, water, earth—and subject to two forces, the centripetal force of love and the centrifugal force of strife. The heavens, he thought, consisted of a crystalline egg-shaped surface to which the stars were attached, with the planets remaining free to move independently.

More important, Empedocles' experiments with a clepsydra—an early timing device—convinced him that air had a corporeal substance. The main body of a clepsydra consisted of a vessel with an opening at the top for pouring in water, and one or more holes at the bottom that allowed the water to drain at an even rate, thereby serving as a clock. Dipping this vessel into a bath, he found that water filled it through the holes on the bottom when the hole at the top was open, but when the top hole was closed, no water would enter because the air in the vessel had no escape. Air, therefore, had to be a substance that needed to be removed before water could take its place.

Empedocles also tried to understand vision and light. Pythagoras had explained vision in terms of particles emitted by the seen object, and others had explained it by postulating that the eye emitted rays that felt the object. (If the latter seems absurd to you, remember that bats "see" objects by emitting sound waves and receiving echoes.) Empedocles reached a compromise solution. For him, the rays emitted by the eye met the emanations from a body.[3] Perhaps his most important conjecture—and that is all it remained—was that light travels with a finite velocity, a speculation not empirically confirmed until more than two thousand years later (not long after René Descartes had declared all of his own philosophy null and void unless light traveled with infinite speed). Legend has Empedocles ending his life while trying to prove his immortality by leaping into the crater of Mount Etna. Witnesses were said to have reported that the volcano contemptuously burped up one of his sandals.

No doubt the most important scientific contribution of fifth-century Greece was the atomic theory, which originated with Leucippos and was further developed in much more detail thirty years later by Democritus. Though little is known about Leucippos—even his birthplace is uncertain—and none of his writings has survived, he is credited with a saying that not only characterizes the science of that time but that may be regarded as the leitmotif of physical science for the next twenty-three centuries: "Nothing happens in vain [without reason], everything has a cause and is the result of necessity." With a few exceptions, physical scientists would subscribe to this credo in one form or another until the nineteenth century, and it would not be decisively abandoned until the twentieth. But that, of course, is jumping way ahead of the story.

Democritus of Abdera, born c. 460 BCE, is the more famous of the two originators of atomism. Abdera, at the northern end of the Aegean, and thus removed from both Ionia and southern Italy as well as Athens, had the reputation, obviously undeserved, of being inhabited by a bunch of dummies. After inheriting a considerable fortune from his father, Democritus spent many years traveling in Egypt, Babylonia, Persia, and possibly even India, searching for and absorbing the knowledge and wisdom of the ancients, which the Greeks believed could be found in the East. Contemporary Egyptians, it seems, looked down on their Greek visitors as childlike newcomers to the world of learning. As an astronomer, Democritus was second-rate: in contrast to the Pythagoreans, he believed the earth to be flat. But he was a good mathematician, who, like Zenon of Elea, thirty years his senior, thought seriously about problems associated with continuity and infinity.

Archimedes later credited Democritus with the discovery that the volumes of a cone and a pyramid are one third of those of a cylinder and prism, respectively, of the same base and height, though exactly how he discovered this fact, later proved by Eudoxos, is not known. However, Democritus's fame rests primarily on his development of the notion of atoms suggested by Leucippos.

Whether the atomic concept was imported from the East or was completely original with Leucippos (who, unlike Democritus, was not widely traveled) is unclear. Similar ideas appeared in Indian philosophy, but four hundred years later. Another tradition ascribes the origin of atomism to the Phoenicians.[4]

In Democritus's view, everything other than the vacuum was made up of atoms—indivisible, eternal, indestructible, infinite in number, and different from one another only in shape, order, and position. Although his theory was completely deterministic, the origin of the atoms' motion remained unexplained. The whole concept was, of course, of an entirely philosophical character and without any empirical support. Nevertheless, in one form or another it remained alive for many centuries. Experimental evidence that would turn it into a real scientific theory had to wait for John Dalton at the beginning of the nineteenth century.

The fifth century BCE produced a number of other significant Greek mathematicians, the greatest of whom was Hippocrates of Chios (not to be confused with the originator of the scientific approach to medicine, Hippocrates of Cos), regarded by some as the father of geometry. He was the first to use letters to refer to geometrical points and to identify lines by the letters designating their endpoints, though in a somewhat more laborious fashion than Euclid did later on. The most famous of the fifth-century Greek philosophers, of course, was Socrates (470–399 BCE), a man of enduring philosophical impact upon Western culture, whose independence of spirit earned him the death penalty for corrupting the youth of Athens. He had little positive effect on science, except perhaps for his skepticism toward the evidence underlying all the scientific speculations of his contemporaries. Let us then turn to the fourth century, dominated by Plato (428–347) and Aristotle (384–322), and Aristotle's prize pupil, Alexander (356–323).

The enormously influential school of philosophy called Academia (located on a piece of land whose original owner was named Academos) was founded by Plato. "Let no one enter here who is ignorant of mathematics" read the inscription over its gate. Plato's philosophy, which perceived reality as a pale image of a world of ideal eternal Forms, made an imprint on the thought processes of mathematicians and other intellectuals that can still be felt today. But his scientific contributions were minimal, and the influence of his thought upon science was not generally fruitful (though, as we shall see at the end of this book, that influence has had a somewhat surprising re-emergence of late). Who discovered that there could be no more than five regular solids, the so-called Platonic solids (the tetrahedron, or pyramid, the cube, the octahedron, with eight triangular sides, the dodecahedron, whose twelve faces are pentagons, and the icosahedron, with twenty triangular faces) is unclear, though Teaitetos of Athens, who studied at Plato's Academy, seems to have been the first to write about them. However, it was Plato who attached great cosmological and metaphysical significance to these five regular polyhedra, all based on fantasy.

The greatest mathematician of that age, Eudoxos of Cnidus (c. 408–355), another pupil of Plato's at the Academy, struggled with the extension of what is meant by a number. This problem—not fully solved until the nineteenth century—arose from the Pythagorean discovery of irrational numbers. (A number is irrational if it cannot be expressed as the ratio of two whole numbers.) His most important substantive mathematical feat was the invention of the "method of exhaustion," a distant precursor of the integral calculus developed by Isaac Newton, and his methodology exerted a strong influence upon Euclid. Eudoxos was also an important astronomer. Not only did he carefully observe the stars, but his application of spherical geometry and the introduction of 27 concentric spheres in order to explain the apparent rotation of the fixed stars, the moon, the sun, and the complicated motions of the planets, as seen from the earth, was the first attempt to understand these motions in mathematical terms. This is why he is regarded by many historians as the real founder of scientific astronomy. However, it was Heracleides of Pontos (c. 388–315) who advanced an astronomical idea that was very far ahead of its time: whereas the earth, he taught, was at the center of the solar system, with the moon, the sun, and the "superior planets" revolving around it and with the "inferior planets," Venus and Mercury, circling the sun, he had the earth rotating daily on its own axis, thereby accounting for the apparent rotation of the body of the stars in the opposite direction. There was not enough observational evidence at the time to convince his contemporaries, but the idea of a rotating earth did not get lost and was repeatedly resurrected (and ignored again) by later astronomers.

Which brings us to Aristotle. Born in 384 BCE in the city of Stageira on the easternmost leg of the three-pronged Macedonian peninsula of Chalcidice at the northern end of the Aegean Sea—his father a physician at the court of the Macedonian king—Aristotle was sent to Athens at the age of seventeen to be educated. There he started out as a disciple of the aging Plato and, except for extensive travel in Asia Minor, remained in Athens for twenty years but seceded from the Academy while his teacher was still alive. When King Philip II of Macedonia needed a tutor for his thirteen-year-old son Alexander, Aristotle moved to Pella and served in that role for three years until the boy had to act as regent of the kingdom in his father's absence. Soon after Alexander succeeded to the throne upon Philip's assassination, Aristotle moved back to Athens to create his own new school, the Lyceum, under the generous patronage of the young Macedonian king, to whom he remained a trusted friend and advisor. The zoo of the Lyceum was stocked largely with animals sent there by Alexander the Great—as the Macedonians (but not the Greeks) called him—during his Asian campaign, and its collection of manuscripts would form the basis of the great library eventually established in Alexandria. After the early death of his royal patron, who had meanwhile conquered a large part of the known world, Aristotle's situation in the anti-Macedonian atmosphere of Athens became perilous, and, to avoid the fate of Socrates, he took refuge in the city of Chalcis in his native Chalcidice, where he died of disease shortly thereafter in 322.

Of Aristotle's many writings, only later edited versions of his notes and lectures have come down to us. These include the *Organon,* a collection of treatises on logic; the *Physica,* on natural science; the *Historia Animalium,* a classification of animals; and *De Caelo,* on astronomy and cosmology.

In contrast to Plato's Academy, the intellectual orientation of the Lyceum was in large part scientific. Perhaps owing to Aristotle's family background in medicine, this scientific disposition leaned heavily in the direction of fact-collecting and experimentation rather than imaginative speculation. But it was not enough for him to record the facts; they had to be explained, and unlike Plato, he was always looking for explanations based in the natural world. Though the phenomena came first, their explanations, if possible, should lend themselves to generalizations and bring forth a theory. Even mathematics was for Aristotle a branch of science about the physical world rather than about a Platonic universe of transcendent ideas.

To understand the workings of nature, Aristotle looked for causes of phenomena, and among these he distinguished four kinds: (1) the efficient cause, which figuratively brings about an effect by a physical effort, (2) the material cause, which furnishes the material of which an object is made, (3) the formal cause, which states the law that underlies an occurrence, and (4) the final cause, which points to its purpose. He looked for mechanisms to gain understanding, and thus for efficient causes, but his scientific explanations often included teleological ideas, that is, final causes. Mechanisms were, in his time, often hard to come by, and teleological explanations were more easily at hand; if the "how" could not be discovered, the "why" would do, and even when the "how" was understood, the "why" was still important.

The shape of the earth, Aristotle agreed with Pythagoras, must be spherical. He based this conclusion on several arguments, the first Platonic and the others empirical: the sphere is the perfect, symmetrical shape; the shadow seen on the moon during an eclipse has a circular edge; and, as you travel north or south, the body of visible stars changes: new stars that had not been seen before rise above the horizon. From the last observation he deduced that the earth cannot be very large, for otherwise that change in the heavens would not be so readily noticeable. In fact, using the shift of the visible stars with a given length of travel, he was able to calculate the diameter of the earth to within about 50 percent of its true value (depending on the length of the unit of measurement, the stadium, which varied from time to time and from place to place).

Seeking to improve on Eudoxos's system of homocentric spheres to account for the motion of the stars and planets, Callippos of Cyzicos (a younger collaborator of Aristotle's at the Lyceum) had increased the number of these spheres to 33, each rotating with its own speed. Aristotle, however, was dissatisfied with this purely mathematical theory, and so he attempted to transform it into a mechanical one by adding 22 new spheres and imagining all of these 55 spheres physically interacting with one another. The result, unfortunately, did not improve the fit of the theory with the data—which consisted of age-old tables assembled by Egyptians and Babylonians, since few of the observations needed for a comparison of these theoretical ideas with reality were made by Greek astronomers. In the later history of astronomy, the prestige attached to Aristotle's views, sometimes conflicting with those of Ptolemy, tended to have a retarding effect on the development of that science.

An important step forward came from Aristotle's notion that astronomy could not be completely divorced from physics—it had to be a part of it. Nevertheless, he divided the universe into two essentially separate regions: the world below the moon and all the remainder. Questions of physics arose primarily in the sublunar sphere, and those of astronomy in the region of the moon and beyond. The sublunar world was distinguished from the heavens by being subject to irregularities and accidents, but even these were not inexplicable: physics too was governed by laws.

Aristotle's laws began with his doctrine of "natural motion." Rejecting atomism and all theories that based physical changes on rearrangements of some basic stuff with fixed characteristics, he regarded the world as made up of five elements: earth, water, fire, air, and ether. The first four permeated the sublunar sphere and the last the translunar region. The natural motion of the former was rectilinear unless blocked, centripetal or downward for earth and water, and centrifugal or upward in the case of fire and air. The downward motion was faster in proportion to an object's weight. For the fifth element, ether, the natural motion was circular and could never be blocked. In the translunar sphere, filled with ether, there existed no bodies in the usual sense, no location, and no time.

The other law of motion was meant to be applicable to ordinary objects in everyday circumstances, such as packhorses or oxen hauling heavy burdens. He said in effect that to move an object of mass M a distance D requires the application of a force F—for Aristotle, that force was essentially the strength required before fatigue sets in—for a time T such that FT was proportional to MD, a proposition that agrees eminently with our experience in a world in which frictional and other resistance to motion are ubiquitous. The magnitude of this resistance was here taken as fixed, but for different friction—motion in water versus motion in air, for example—the constant of proportionality (though he does not put it that way) between FT and MD was assumed to increase with this friction. Aristotle believed that if it were possible to produce a vacuum, which would enable objects to move without friction, any force would move an object with infinite speed, from which he drew the important conclusion that in the sublunar region a vacuum had to be physically impossible.

Since D/T is the distance traveled divided by the time it took, that ratio is the average velocity V of the moving object; therefore, the Aristotelian law of motion was later interpreted as decreeing that the required force F is proportional to the product MV of the mass of the moving object times its velocity, and it was considered valid until abrogated by Isaac Newton. It should be remembered, though, that Aristotle did not intend his law of motion to be taken out of its context of earthbound experience, and it did not occur to him to apply it to objects such as the planets in the translunar world.

Aristotle's writings—which were translated into Latin and Arabic in ancient times, and from Arabic back into Latin during the medieval period, after the original Latin translations were lost—had a profound impact on Western thought and beyond. In the Middle Ages, the Catholic Church became actively hostile to his philosophy, to the point of prohibiting the reading of his works, a proscription that turned out to expand their study. Healthy and stimulating at first, eventually Aristotle's scientific doctrines became stultifying because of the rigidity with which they were applied. His sway did not end until the scientific renaissance brought about by Galileo and Isaac Newton. However, even the later development of physics showed the lingering influence of Aristotle in the continuing primacy of the search for explanations of the way things change and move, with less emphasis on Platonic elucidations of the way they *are.* Only in the second half of the twentieth century would Plato's way come back into vogue.

The end of Alexander's conquest marked the beginning of the Hellenistic age (a period when Greek culture dominated the Mediterranean and whose name derived from the word *hellene,* which the Greeks used to describe themselves), and Euclid was its first great mathematician. His singular work, the *Elements of Geometry,* made an indelible imprint for well over two millennia. Little is known about the man or his life, not even the dates and places of his birth and death. He probably received his mathematical education in Athens at the Academy and spent most of his life in Alexandria, the great city founded by Alexander after his conquest of Egypt.

Euclid's *Elements* consists of thirteen books that have come down to us in their entirety. The first six deal with plane geometry, including much of algebra viewed from a geometric perspective; books seven through ten address arithmetic and the theory of numbers; and books eleven through thirteen take up solid geometry. The importance of this work rests on two characteristics: the beauty and power of the methods employed and the richness of the propositions announced and proved. The influence of both Euclid's general methodology and his geometrical approach to algebra would still be expressed in Newton's *Principia* (though this particular aspect of Euclid's influence makes the *Principia* very hard to read; Euclid's geometrical approach to algebra is no longer fashionable).

The clean and minimalist elegance of the organization and presentation of the *Elements*, in which propositions (now called theorems) are clearly stated and then proved and followed by conclusions, their order dictated by their logical succession, is proudly imitated by mathematicians to this day. Definitions and axioms (basic assumptions that are made in logical reasoning) are followed by a set of postulates (basic mathematical propositions that are not susceptible to proof but have to be accepted; today, these are usually also called axioms). From these definitions, axioms, and postulates, all the theorems follow and all geometrical problems can be solved by logical reasoning without the need for any additional, extraneous assumptions. Aristotle's influence led Euclid to accept the need for such postulates but also to reduce their number to its absolute minimum.

This "axiomatic method," which had been slowly evolving for some time before Euclid became its prime expositor, continues to have a profound impact on the way in which mathematicians formulate and communicate the results of their work, sometimes even influencing scientists in other fields. What is more, Euclid's particular choice of geometric postulates has shaped the modern development of this field of mathematics, and of physics as well.

One postulate, in particular, continued to stick in the craw of everyone who seriously thought about geometry: the fifth and last one, which he stated this way: "If a straight line crossing two straight lines makes the sum of the interior angles on the same side less than two right angles, the two straight lines, if continued indefinitely will meet on that side on which the two angles add up to less than two right angles" (Fig. 4).

This statement seems so obvious that almost everyone will accept it and may see no need even to mention it explicitly. However, ever since Euclid's day, mathematicians have tried to do away with it as a postulate, either by replacing it without disturbing the rest of geometry, or else by relying on Euclid's other four postulates and proving the fifth as a theorem. Replacing it by another without ill effect on geometry turned out to be easy enough, though to little advantage. Several substitutes were proposed, an example of which is "Through a given point only one parallel can be drawn to a given straight line," but they all turned out to be essentially equivalent to Euclid's. (This is why the fifth postulate is often stated in the form given in the last sentence and referred to as the "parallel postulate.")

A proof that the fifth postulate could not be deduced as a theorem from the other four had to await the great nineteenth-century mathematician Karl Friedrich Gauss. The work of other nineteenth-century mathematicians demonstrated that it could be replaced by a

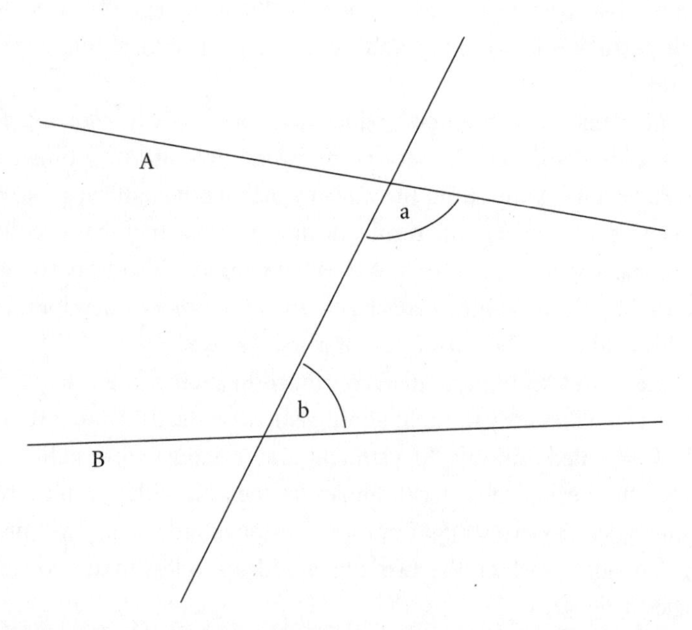

Figure 4 Euclid's fifth postulate states that if the angles a and b add up to less than 180°, the two lines A and B must eventually meet on the right.

different, contradictory postulate, thereby constructing alternative geometries known as non-Euclidean. These turned out to have a profound effect on physics and on our conception of the structure of the universe.

An anecdote illustrates why Euclid may be regarded as the founder of what we now call basic science. When one of his students asked him what it would profit him to learn all the things Euclid taught, the teacher gave him an obol (a Greek coin), so that he might gain from what he learns. The basic purpose of science is not its application and profit, but understanding what makes nature tick.

Science in Hellenistic Greece continued with the astronomers Aristarchus (c. 320–250) and Hipparchus (second century BCE). Born on the island of Samos off the Ionian coast near Miletos, Aristarchus contributed two important sets of ideas. The first, contained in his preserved treatise *On the Sizes and Distances of the Sun and Moon,* was to use what amounted to trigonometry (though trigonometry as such did not exist) to estimate the sizes of the sun and moon, albeit using grossly deficient observational data and therefore arriving at results that were highly inaccurate. Nevertheless, his trigonometric methods represented a decisive step forward.

For example, at the time of a half moon, when our line of sight to the moon and the line from the moon to the sun must make a right angle (so that we see the moon lit up exactly from the side), he measured the angle between the lines of sight to the moon and the sun (Fig. 5). From his result of 87° he was able to conclude that the distance from earth to the sun was nineteen times that from the earth to the moon. Had he used the correct value of 89° 50′ for that angle, he would have arrived at the correct ratio, which is about 400.

However, his most important work (with which we are familiar only because of a report by Archimedes, a younger contemporary) described Aristarchus's view of the universe. He regarded the sun and the fixed stars as unmoving, the earth and the planets as circling the sun, the moon as circling the earth, and, to account for the apparent rotation of the fixed stars in the heavens, the earth as revolving on its own axis, just as Heracleides had it. The size of the cosmos, he believed, was very much larger than had been thought before, the distance to the fixed stars so large that the orbit of the earth around the sun was, in comparison, like a point. Thus, Aristarchus roughly anticipated Copernicus and modern astronomy by eighteen centuries. Moreover, he did this not through wild speculation but through careful astronomical reasoning. Since his observations, faulty though they were in detail, had led him to the correct conclusion that the sun is very much larger than the earth, he could not believe that it would be the larger sun that revolved around the smaller earth; it had to be the other way around. For good reason, Aristarchus has been called "the Copernicus of antiquity." Removing the earth from the center of the universe earned him the same accusations of impiety that Copernicus would suffer later on. Though his views were not completely lost, they were more or less ignored. (An exception was the Babylonian astronomer Seleucus, who lived in the early part of the second century BCE, probably in the city of Seleuceia on the Tigris.)

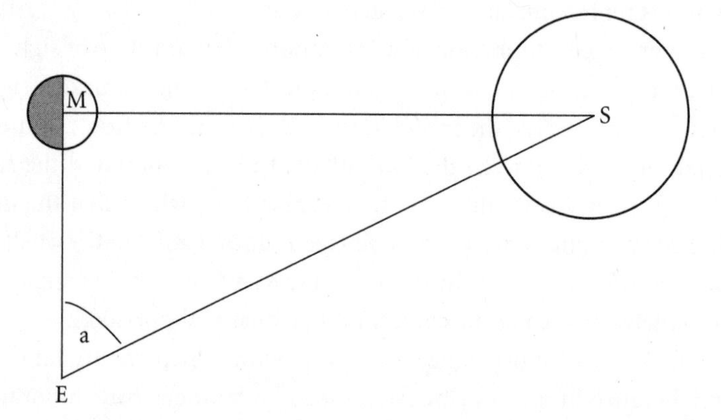

Figure 5 Since the moon is seen from E exactly half lit, the angle at M is 90°. By measuring the angle *a* between the lines of sight to the moon and the sun, Aristarchus was able to calculate the ratio ES/EM.

At this point we might want to take another look at China, on the other side of the world, to see what contemporary mathematicians and astronomers there were doing. The substance of the greatest arithmetical Chinese classic, *Chiu-chang suan-shu,* or *Arithmetic in Nine Sections,* whose original date is uncertain—tradition ascribes it to the twenty-seventh century BCE—has come down to us via the mathematician Chang Ts'ang, who died in 152 BCE at an age of over 100. Revised and enlarged in the next century by Keng Shou-ch'ang, minister to the emperor Hsuan-Ti, it contains measurements of plane and solid geometrical figures, extraction of square and cube roots, linear equations with several unknowns, problems involving quadratic equations, the earliest known mention of negative quantities since the Babylonians, and the Pythagorean theorem; the value of π in it is 3. In the third century CE, the mathematician Wang Fan would improve the value of π to 3.155.

Concerning astronomy, the authors of a very instructive little book, *The Way and the Word: Science and Medicine in Early China and Greece,* tell us that the traditional Chinese view of the beginnings of astronomy comes from the *Institutions of the Emperor Yao,* written in the second half of the fourth century BCE, though claimed by tradition to be much older. The emperor "commanded the Hsi and Ho [families of hereditary astronomers] reverently to follow august heaven, calculating and delineating the sun, moon, and other celestial bodies in order respectfully to grant the seasons to the people." After calculating a reliable calendar, "if you earnestly supervise all your functionaries, your achievements will be resplendent."[5] The study of the heavens and the earth, in other words, was done at the command of the ruler and for the direct benefit of his subjects. Note the contrast between this command performance for the benefit of the people and the individual, curiosity-driven research for understanding by the Greek scientist, to whom we now return.

We find that even the greatest observational astronomer of antiquity, Hipparchus, regressed to a geocentric model for the universe,

Aristarchus's insights notwithstanding. In order to account for his vast set of assembled precise data with a minimum of hypotheses, he stuck to the traditional view of the earth at the center of the universe. Little is known about the life of Hipparchus, except that he was born in Nicea, now called Iznik, in northern Asia Minor just east of the Sea of Marmara, and that he did most of his observational work on the island of Rhodes, using a number of instruments of his own invention. Listing the positions of about 850 stars ordered by their magnitude, he assembled the first-ever star catalogue, employing a system that resembled the one in use today. So accurate was his work, and so extensive, that it was not only taken over *in toto* by Ptolemy (which led to subsequent accusations of plagiarism against the latter) but was used even by Edmund Halley at the end of the seventeenth century.

Ptolemy, or Claudius Ptolemaeus, was of course the famous Egyptian astronomer, astrologer, and philosopher of the second century CE whose model of the solar system exerted an enormous influence on European science—and even religion—for many centuries to come. Except for the fact that he lived and worked mostly in Alexandria, which had by then fallen under Roman domination, very little is known about his life. The *Almagest* (its Arab name) contained Ptolemy's model for the motion of the moon, sun, planets, and stars, all orbiting the earth. The essence of his geocentric system—like Hipparchus, he abandoned the heliocentric model Aristarchus had proposed—was strongly influenced by the views of Plato: as circles were the perfect shapes, all the motions of the heavenly bodies had to be uniform and basically circular. How, then, to account for the observed periodically retrograde motion of the planets? This he explained by a system of epicycles: the trajectory of each planet consisted of a circle whose center, in turn, uniformly moved along a larger circle centered at the earth. The resulting planetary motion was therefore not circular, and as seen from the earth—the center—there were times when a planet appeared to be moving backward.

Complicated as the Ptolemaic model was, it did not really account for the observed data very well, but European civilization stuck with it for some 1,300 years.

Meanwhile, we must not forget the greatest of the Greek scientists, Archimedes, who lived just before Hipparchus. Son of the astronomer Pheidias, Archimedes was born in 287 BCE in the Greek outpost of Syracuse on the Carthaginian-dominated island of Sicily, but also spent some time in Alexandria, then the center of the scientific world. When the Romans besieged Syracuse in 212, Archimedes was reported to have assisted the defense of the city by a variety of machines of his own invention, including catapults and an arrangement of concave mirrors whose reflected sunlight set Roman ships on fire. Legend had Archimedes slain by a Roman soldier during the sack of Syracuse as he was drawing geometrical figures in the sand, unwilling to be interrupted. His tomb, neglected for many years, was restored and the tombstone described by Cicero in 75 BCE, but its location is now unknown.

Archimedes was that rare combination, an accomplished mathematician and scientist and at the same time an inventor with an extremely fertile imagination for practical machinery. His most original mathematical work was in geometry, including proofs of the formulas for the volume and surface of a sphere ($4/3\ \pi r^3$ and $4\pi r^2$, respectively). He was so proud of these proofs that he ordered an image of a cylinder circumscribing a sphere to be placed on his tomb after his death. Archimedes did not use equations; the formulas for the volume and surface area of the sphere were expressed as ratios to the corresponding figures for the cylinder that circumscribed a sphere (hence the image for the tombstone, now unfortunately lost). The remarkable thing is that these results were obtained without the use of calculus, of which Archimedes is regarded as a precursor. In order to find a good value for the number π, he calculated the areas of two regular 96-sided polygons respectively inscribed in and circumscribing a circle, which resulted in the inequalities $(3 + 1/7)$

$> \pi >$ (3 + 10/71), or (in modern decimal notation) $3.142 > \pi > 3.141$.

The Archimedean spiral bears his name to this day (Fig. 6). By his definition, it is formed "if a straight line of which one extremity remains fixed, be made to revolve at a uniform rate in a plane until it returns to the position from which it started, and if, at the same time as the straight line revolves, a point moves at a uniform rate along the straight line, starting from the fixed extremity, the point will describe a spiral." (In modern language, on any ray, the distance from the center is proportional to the angle the ray makes with a fixed line: $r = \alpha\theta$.) Though severely hampered by the lack of a useful notation system, Archimedes also invented expressions for extremely large numbers (such as his estimate of the number of grains of sand that would fill the universe), a facility that lay dormant for many centuries to come.

In physics, Archimedes made contributions to astronomy and optics, but his work was primarily concerned with the parts of mechanics known today as statics and hydrostatics, of which he may be re-

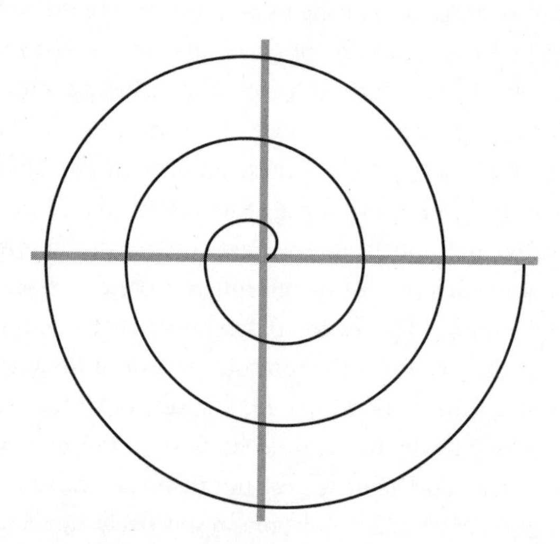

Figure 6 An Archimedean spiral.

garded as the founder. Writing in a Euclidean style, beginning with definitions and postulates, his approach to physical problems made him the first mathematical physicist that we know of. For example, after stating two postulates involving weights in equilibrium, he proves that "two magnitudes, whether commensurable or not, balance at distances reciprocally proportional to them," on the basis of which he solves the problem of "how to move a given weight by a given force." His solution induced him to make the legendary announcement, "Give me a point of support and I shall move the world." To demonstrate to the king that this was not an empty boast, he lifted an entire ship, with its crew and freight, by means of a compound pulley.

In hydrostatics he discovered what is still known as the Archimedean principle: the weight of a body wholly or partially immersed in a fluid is reduced by an amount equal to the weight of the fluid that the body displaces. According to legend, after noticing the lightness of his body in his bath and suddenly realizing why this was so, he ran naked through the streets of Syracuse shouting *Eureka.* The principle leads to an easy method of detecting when an object (such as a king's crown) is made of pure gold or is adulterated by an admixture of lighter metals: if its weight in air is W and its weight when immersed in water is w, the ratio of the weight in air to the difference between the two weights, $W/(W - w)$, should be the same as the corresponding ratio for any other object known to be made of pure gold, no matter its size or shape. If the ratio is less, it is not pure. That ratio is the average specific weight of the material that constitutes it, if the specific weight of water is taken to be 1.

If the theoretical and mathematical work of Archimedes was forgotten for many years, not so his legends and practical inventions. Perhaps the best known of the latter is the Archimedes screw (Fig. 7), which is still used in modern factories to move powdery substances. It consists of a circular cylinder within which a tight-fitting spiral screw can be turned by a handle at one end. If the lower end of the ingenious device, tilted upward, is dipped into a liquid and the spiral is turned, it will gradually be filled and spill its contents at the upper end.

It is interesting again to glance toward China and to contrast the ease with which the Greeks accepted new technology and new practical inventions such as those of Archimedes with the dim view of technical innovation taken by the Chinese at the time. The prevailing contemporary philosophy toward science, Taoism, was not hostile to science as such. But, new inventions that might make certain burdensome tasks easier were viewed with deep suspicion. A good example is given in the Taoist text *Chuang Tzu,* where a farmer is shown a new device, called the swape, that would make irrigation of his field much easier. "I have heard from my master," he responds, "that those who have cunning devices use cunning in their affairs, and those who use cunning in their affairs have cunning hearts. Such cunning means

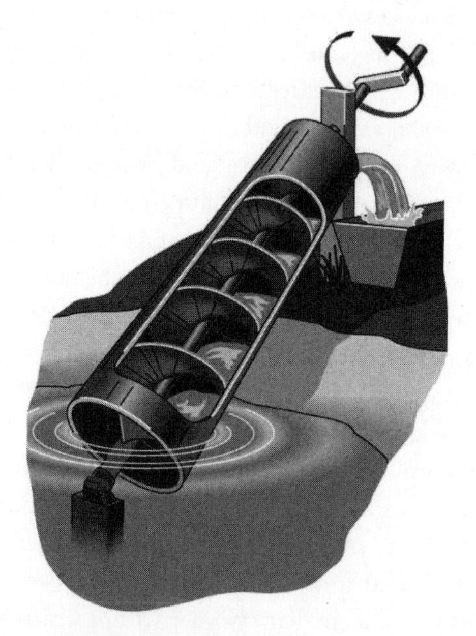

Figure 7 Archimedes' screw in action.

the loss of pure simplicity. Such a loss leads to restlessness of the spirit, and with such men the Tao will not dwell. I knew all about [the swape], but I would be ashamed to use it."[6] Needham's explanation of this suspicion is the Taoists' generalized complaint against their contemporary society, which became a hatred of all "artificiality" and their view that technical inventions had contributed to the differentiation of classes.

The last of the great Greek mathematicians, Apollonius, was born in about 260 BCE in the town of Perga on the south coast of Asia Minor, which at the time belonged to the kingdom of Pergamon. He studied at the school founded by Euclid in Alexandria and continued to flourish there, but very little is known about the rest of his life. His major legacy is his work, contained in an eight-volume treatise, on conic sections—the curves formed by cutting a circular cone with a plane at various angles: circles, ellipses, parabolas, and hyperbolas. (The names of all but the first of these were coined by Apollonius.) These particular curves turned out later to have great significance in dynamics and in studying the trajectories of projectiles and the orbits of planets as well as comets.

Let us, then, summarize the state of physical science in the West at the waning of the classical Greek and Hellenistic civilizations and the rise of the Roman empire and the Christian era. Scientists—this word is an anachronism, of course, since nowhere in the world did they call themselves by such a name—were no longer satisfied just knowing facts and tricks of the trade; they were searching for understanding and explanations, preferably in terms of mechanical or teleological causes.

In the Greek realm, explanations, especially in astronomy, were often formulated in mathematical terms. Competing theories of the constitution of matter and the structure of the universe were, for the most part, based on nothing but philosophical speculation, but some used sound reasoning to draw valid conclusions, albeit incorrect in their details owing to poor observational data. Records of detailed observations of the celestial bodies existed in the Hellenistic world as well as in China and India. Several different models of the solar system and the cosmos were put forward, but the one closest to the modern view, the heliocentric model of Aristarchus, was generally ignored in favor of a complicated geocentric construction by Ptolemy.

In mathematics, important progress had been made in geometry, algebra, arithmetic, and number theory, but except for the field of geometry, the most advanced formulations of mathematical ideas by Greeks were still severely impeded by a very clumsy and primitive notation. The crucial step forward was the concept that it was not sufficient to treat each individual mathematical problem by itself as it arises, but to prove theorems that would encompass the solutions for whole classes of problems, irrespective of their contexts. The quintessentially Greek notion of an airtight logical proof of a theorem turned out to be of seminal importance not only to assure correctness of results, but also for the further progress of mathematics. The most valuable proofs opened up entirely new vistas of research—and it remains so to this day. Though they contributed significant mathematical results, Chinese and Indian thinkers never came up with this idea, much to the detriment of mathematical developments in their cultures.

Compared with the state of physical science in the world some seven hundred years earlier, the Greeks had made enormous advances, even though their progress was based in large part on earlier Babylonian and Egyptian traditions. With that we turn to the story as it developed after Christianity slowly began to dominate Europe, with the help of the Roman empire. At the beginning of the Christian era, the overarching conviction carried by scientifically knowledgeable philosophers was that nature was orderly, subject to regular rules, and deterministic. This was perhaps the most significant scientific legacy of classical Greece.

three

Science in the Middle Ages

As the power of Rome gradually rose, interest in philosophy and science, even among the Greek-speaking and Greek-educated intellectual elite, declined. The empire created by Roman military might produced vast technical improvements such as a far-flung network of roads and aqueducts, as well as cities with paved streets and plumbing facilities. But its citizens, including the most educated among them, had less interest in abstract or theoretical matters. As a result, even though during the early centuries of the Roman republic there were still a number of good Greek scientists, mysticism and astrology flourished while progress in science languished, starved by want of an intellectual milieu to nurture rational creativity. For a prolonged period, not only did original scientific thinking wane, but eventually parts of the old knowledge were lost. This was true particularly in the western part of the later empire, where a steadily dwindling class of the population read Greek, without which they had no access to the works of the classical mathematicians, astronomers, and natural philosophers.

One of the few original contributors to the propagation of classical thought was the Epicurean poet Titus Lucretius Carus, who probably lived from 99 to 44 BCE. Almost nothing is known about his life, other than calumnies written about him some 475 years later by St. Jerome, who disliked him because of his anti-religious views. His *De rerum natura,* an epic treatise written in Latin dactylic hexameters and running almost as long as the *Aeneid,* is a work of scientific philosophy dealing with atomic physics and cosmology. For Lucretius, nothing existed in the infinite universe but indestructible atoms of various shapes, without color, sound, taste, odor, or temperature, moving irregularly in empty space.

As a substitute for original thinking, writers in subsequent centuries turned to the compilation of scientific handbooks and encyclopedias, which, in order to be accessible to those without the Greek tongue, appeared, not always adequately, in Latin. The greatest of the early Latin encyclopedists, Pliny the Elder (23–79 CE), was a Roman cavalry commander who retired from service under the emperor Nero to devote himself to a compilation of all the known sciences of the day, along with much history and grammar, and who died in the eruption of Mount Vesuvius that inundated the cities of Pompeii and Herculaneum. His *Natural History* served as the main source of scientific information for the next thousand years. Three of the best of the late antique encyclopedists, who continued to exert considerable influence throughout the early Middle Ages, were the Roman scholar and statesman Ancius Manlius Severinus Boethius (c. 480–524), who was executed for treason by the Ostrogoth King Theodoric, whom he had served in Ravenna; St. Isidore (c. 560–636), bishop of Seville; and the English theologian Venerable Bede (c. 673–735).

The greatest mathematician of the era, Diophantus, who probably flourished during the late third century, made important contributions to number theory, algebra, and the theory of equations (one kind of which still bears his name). But other than Diophantus and Ptolemy, the only thinkers about physical science during the first millennium of the Christian era worth mentioning were Hero of Alexandria, who lived in the first century CE, before the rapid growth of Christianity and the establishment of the Christian Church, and

John Philoponus (John the Grammarian), who flourished in the sixth century.

Still the center of science in the Greek and Roman world, Alexandria was an appropriate place for Hero to set up a technical school with a strong emphasis on research. His most notable contributions were experiments with steam pressure, which led him to recognize that air was an elastic medium, capable of compression and expansion. Though regarded for a long time as a third-rate tinkerer rather than a serious scientist, Hero's conclusion that air consisted of minute moving particles was some fifteen centuries ahead of its time. He also contributed to the science of mechanics and to mathematics, especially in regard to geometric measurements.

The importance of the Christian neoplatonist philosopher Philoponus, who also lived in Alexandria, derives from his extensive critical and original commentaries on Aristotle. Based on experiments, and anticipating the concept of inertia, he refuted Aristotle's doctrine that heavier bodies fall faster than lighter ones as well as his stricture against the possibility of a vacuum. His ideas, though they put him a whole millennium ahead of his time, remained essentially unknown in the Latin West during that entire period.

After the emperor Constantine became a Christian early in the fourth century, Christianity spread to the far reaches of the crumbling western empire, including the areas populated by Germanic tribes. The preservation of the remnants of the scientific tradition that had come down from the Greek natural philosophers now depended almost entirely on the attitude of the Church. Fortunately, after strong initial hostility, the Christians' view of the pagan Greek knowledge that many of them (including St. Augustine, 354–430) had learned in their youth was cautiously accepting, certain specific objections notwithstanding. As St. Clement (c. 150–215) and his disciple, the theologian Origen of Alexandria (c. 185–254), taught, whether Greek philosophy was good or bad depended entirely on how Christians used it. Pagan astronomy and philosophy, including natural philosophy, if studied warily—and, if necessary, reinterpreted—could help in preparing for the study of the Bible and improving one's deeper understanding of it. The view that pre-Christian learning could thus serve a useful purpose as handmaiden to Christian theology much facilitated the further development of science in medieval Europe.[1] Later, the Church would express antagonism toward certain scientific discoveries, such as the Copernican heliocentric model of the solar system and Darwinian evolution. But the development of Western science could have been seriously impeded, if not wholly foreclosed, had Christianity in the Middle Ages adopted a more fundamentally hostile stance toward the pagan philosophical ideas Europe inherited from the Greek civilization and toward their off-spring, which eventually grew into modern science.

With the collapse of the western Roman empire and the shrinking of its cities, one of the major benefits of civilization—the education of an elite (at that time meaning mostly the future clergy)—became confined first to rural monasteries and later to cathedral schools, where it was pursued entirely in Latin. A number of teachers at the cathedral schools in major cities became widely known. They included the Benedictine monk Gerbert (c. 946–1003), sometimes credited with the invention of the escapement, a crucial element in clocks, who had a wide knowledge of astronomy and later became Pope Sylvester II; and the French philosopher Peter Abelard (1079–1142), who attracted large numbers of students. The subject matter of their instructions tended to be quite broad, including as much science and mathematics—which was usually relatively elementary—as they could find in the classical Latin literature and in Latin translations of Arabic treatises imported via Spain. Few of the great Greek works on natural philosophy or mathematics were directly translated into Latin until the thirteenth century, which is why the Latin encyclopedias and the new influx of science and natural philosophy coming from the Arabs became crucial sources of information.

The year 622 saw the birth of Islam and the beginning of the rapid

Arab conquest of the area from India, across northern Africa, and into Spain. As Christian Western Europe turned inward, the expanding Muslim world experienced a great flowering of intellectual activity. Caliph Harun-al-Rashid, a contemporary of the Holy Roman Emperor Charlemagne, actively encouraged the translation of all the treasures of knowledge to be found in Greek into the Arabic language, as had the earlier caliph al-Mansur, the founder of Baghdad. During the ninth century, almost every significant mathematician was also an astronomer or astrologer. And in this part of the world, including Islamic Europe, they were, with a few Jewish and Christian exceptions, Muslims, all speaking and writing in Arabic. Arab culture assimilated Greek knowledge, and most of the existing Greek manuscripts, including the texts of Aristotle, Plato, Euclid, and Ptolemy, were translated into Arabic and Syriac, thereby greatly assisting their later survival in Western Europe.

Islamic intellectuals and scholars were primarily supported by secular patrons; in contrast with their Christian counterparts, they were not clerics, which allowed their thought more leeway. On the other hand, this freedom from religious affiliation subsequently exposed them to much more ferocious suppression by clerical authorities, which may go a long way in explaining why the exuberant flowering of Islamic culture in the Middle Ages did not spawn a scientific revolution comparable to that in Christian Europe.[2]

The most influential of the ninth-century Arab mathematicians (and an astronomer as well) was al-Khwarismi, a corruption of whose name eventually produced the word *algorithm* for any systematic rule of calculation. (Similarly, the word *algebra* originated from a corruption of the title *Al-jabr,* his work based on earlier achievements of Diophantus and of the great Indian mathematician and astronomer Brahmagupta of the seventh century, which gave an exhaustive exposition of linear and quadratic equations.) He was the first to expound, systematically, the Hindu number system (without any claim that it was his own invention). The transfer of this system from India to the Arabs and thence in the twelfth century to Western Europe (where these numbers became known as Arabic numerals) was of seminal importance to the subsequent scientific development of Western culture.

The decimal place-value number system—the system in use everywhere today, in which 3756 stands for $(6 \times 10^0) + (5 \times 10^1) + (7 \times 10^2) + (3 \times 10^3)$—had been evolving in India for several centuries before its appearance in the influential versified work *Aryabhatiya* by Aryabhata in 499 CE. It still took some time after that for the system to develop to the point where only nine symbols were used to denote any number, no matter how large. Even then it was still the same as the place-value system employed much earlier by the Sumerians—since no symbol for zero existed as yet—and it may have been imported from there. The first documented appearance of a zero is in an inscription dated 876. Of the three elements of the so-called Arabic numeral system—the decimal base, positional notation, and a simple symbol for each of the ten numerals, including zero—none was really original with the Hindus; they all were much older. But the Hindus apparently were the first to combine all three, and the Arabs, via Al-Khwarismi, learned the system from them. (A positional number system was also independently invented in the Western Hemisphere by the Mayas, whose civilization flourished from the third to the tenth century.)

Another ninth-century Arab mathematician of note, Thabit ibn-Qurra, founded a school of scribes who performed the invaluable job of translating into Arabic many of the works of Euclid, Archimedes, Apollonius, and Ptolemy in the second half of the century. But equally important, he also commented on these Greek works and suggested valuable generalizations and modifications. Thabit ibn-Qurra and al-Battani (known in Europe as Albategnius), who was primarily an astronomer and lived well into the tenth century, both further developed the trigonometry that had come down to them from the Greeks and the Hindus. Whereas Ptolemy had employed for his astronomical data the relation between the length of a chord of a circle to the angle subtended at the center, the *Surya Siddhanta*—written about 400 CE, though there are variously dated versions of it—had used for the first time the relation between the half-chord and the corresponding half-angle, thereby effectively introducing the sine function. (The word *sine* derives from a mistranslation of the Arab version of the Hindu word *jiva*.)

The greatest of the Muslim natural philosophers, Ibn al-Haitham, known in the Latin West as Alhazen, was born c. 965 in Basra but flourished in Egypt and died about 1039 in Cairo. In addition to commentaries on Aristotle and Galen, he wrote a work on optics, *Kitab al-manazir,* which dealt with the properties of spherical and parabolic mirrors, spherical aberration, refraction, the magnifying power of lenses, optical illusions, properties of the eye, and the rainbow. A translation of this work into Latin continued to exert a powerful influence on Western science for many centuries to come. Alhazen also had a correct explanation for the apparent increase of the size of the sun and the moon as they approached the horizon, and he was the first to explain in detail the principle of a camera obscura. (A camera obscura is a dark room or box with a small hole on one side through which light shines onto the opposite wall, projecting an inverted image of the outside scene. It had been in use in China as early as the fifth century BCE, and Aristotle had noticed the inverted crescent image of a partially eclipsed sun projected on the ground through small gaps in tree foliage. After the invention of photographic emulsion, small versions of the camera obscura, with a photographic plate or film placed opposite its pin hole, became the popular box camera.)

Twelfth-century Christian Europe underwent a decided intellectual shift in its view of how God governed the world, though this subject continued to remain controversial for several hundred years. Rather than determining every occurrence in nature, God came to be seen as responsible for setting up nature in such a way that chains of events caused other chains of events, and natural philosophy became the appropriate discipline for the study of these phenomena of the world. Theological objections notwithstanding, the universe was once again viewed in the way that many of the Greek natural philosophers had seen it: as a self-operating, regularly functioning machine, created and set in motion by the hand of God but running on its own. William of Conches, a philosopher in the School of Chartres and one-time tutor of the young Henry II, defiantly declared that "the reason behind everything should be sought out," and he was not alone.[3] It was important for Christians to know the laws of nature in order to appreciate the sublime masterpiece God had created.

This curiosity about nature among Western European scholars was sparked in large part by their discovery of a treasure trove of works by Greek mathematicians and natural philosophers, Aristotle above all, in the libraries of Toledo, Segovia, and Cordoba. By the twelfth century, Spain had been in the hands of the Moors for some five hundred years, but its Christian reconquest was well under way. Among the treasures the Christians claimed were the writings of Arab mathematicians as well as classical works, some in the original Greek, others in Arabic translations, accompanied with elucidating commentaries by such Islamic scholars as Averroes (Ibn Rushd, 1126–1198), Jewish sages such as Moses Maimonides (1135–1204), and others.

These discoveries created among the comparatively ignorant Christians of Europe an insatiable thirst for the knowledge of the much more civilized pagans and infidels, a thirst that could be slaked only by translating all the old fonts of wisdom into Latin so as to make them readily available to Western scholars. Great translation centers grew up both in Spain and in Sicily—which the Arabs had wrested from the Byzantine empire and the Normans had subsequently conquered. The influx of first- and second-hand ancient Greek as well as more recent Arab and Jewish learning rekindled embers that eventually sparked the Renaissance and its scientific revolution.

As cities started to regain their vitality in Western Europe, the twelfth century saw the establishment of the first European universities. Paris and Oxford in particular played important roles in physical science and natural philosophy, while Bologna confined itself to law and medicine. The difference between universities in Christian Europe and their analogues, the madrasas in the Islamic world, is noteworthy.[4] The latter, set up on the basis of private endowments, were exclusively for the propagation of religious knowledge and Islamic law; philosophy or the natural sciences, the so-called foreign sciences, were never a focus of attention there, particularly because Aristotle's teachings were sharply at variance with those of the Quran. The foreign sciences were available in Arabic libraries (though occasionally they were burned), but, much as some prominent independent Arabic thinkers admired them, they were rarely taught to madrasa students.

The influence of the translated works of Aristotle at that time is impossible to overstate; the following encomium by Averroes is an example of the medieval opinion held by Islamic and Christian thinkers alike: "The teaching of Aristotle is the supreme truth, because his mind was the final expression of the human mind. Wherefore it has been well said that he was created and given to us by divine providence that we might know all that is to be known. Let us praise God, who set this man apart from all others in perfection, and made him approach very near to the highest dignity humanity can attain."[5] Aristotle's teachings were taken as gospel—and inevitably this led to a clash with the guardians of the real gospels.

The place where, in the thirteenth century, the conflict was especially fought out was the University of Paris, whose most important house of learning was founded in 1257 by the theologian Robert Sorbon and is associated with his name to this day. The strife at the university between Aristotelians and the conservative powers of the Church led to an official ban of Aristotle's books dealing with natural philosophy that stayed in effect for about forty years, though it did not extend to other universities such as Oxford. Even after the ban was lifted, the struggle between the faculties of theology, where of course faith reigned supreme, and arts, where the works of "the philosopher" were revered, continued until the bishop of Paris in 1277 issued an edict explicitly condemning 219 of Aristotle's propositions. At least some Church officials had given up on the idea that natural philosophy could serve as the handmaiden of theology. To demonstrate which was supreme, faith or reason, the arts masters at Paris were obliged, by oath, to resolve any conflict arising between the two in favor of faith.

The theologians objected most vociferously to three of Aristotle's propositions: the eternity of the world, the doctrine of the double truth, and the limits on God's power. "The world as a whole was not generated," Aristotle taught, "and cannot be destroyed, as some allege, but is unique and eternal, having no beginning or end of its whole life."[6] This clearly was in conflict with the account of creation by Genesis. How did Christian scholars deal with such contradictions between philosophy, based on reason, and theology, based on faith? As the thirteenth-century Scandinavian, Boethius of Dacia, put it, "Who does not believe these things is a heretic; whoever seeks to know these things by reason is a fool."[7] Faith and reason simply operate on different planes and are incommensurate. While conservative theologians objected, the Dominican friar Thomas Aquinas (1224–1274) agreed: "That the world had a beginning . . . is an object of faith, but not of demonstration or science."[8] The world was created by God, and yet it might, nevertheless, be eternal. What is more, Aquinas preached, studying the laws governing this world allows us to discern God's intentions in creating it, a view strongly opposed by the Scottish Franciscan friar and Aristotelian philosopher John Duns Scotus (c. 1266–1308).

Thus, medieval Western Europe managed to find a compromise that prevented Christian faith from blocking reason altogether and made the later rise of science possible. The price was an implied doctrine—never acknowledged and in fact explicitly condemned by the edict of 1277—of "the double truth," which allowed the truth of religion to coexist with the truth of science or reason.[9]

The third and perhaps most important issue dealt with in the Paris edict was God's absolute power. The question was whether God could do things that contradicted the laws of nature (the specific laws enunciated by Aristotle, to be sure, but the same problem would arise later with respect to other natural laws). Aristotle had declared that a vacuum was impossible. Did that mean God did not have the power to make a vacuum? Every occurrence had a cause, according to Aristotle; so God could not produce any occurrence he wished, natural cause or no? It was, in essence, the question whether miracles could happen, and of course the Church could not tolerate a natural philosophy that denied their existence. The bishop of Paris effectively decreed that God had the power to do anything he desired, lawful or not. That Christians were obliged to obey the bishop's decree caused complications for science, but these proved surmountable.

Christian Western Europe did produce a certain number of creditable physical scientists and mathematicians in the thirteenth century. The greatest of these was the Italian Leonardo Fibonacci (or Leonardo Pisano, c. 1170–1240), who was the first to give a complete explanation of the Hindu numerals as well as of Hindu and Arabic arithmetic to the Christians. His use of algebra to solve geometrical problems was a novelty in medieval Europe, but today Fibonacci is primarily known for the sequence 1,2,3,5,8,13,21, . . . (each term of which is the sum of the two preceding terms), which turned out to have applications in biology.

Others of note were the German mathematician and physical scientist Jordanus de Nemore (Nemorarius), thought by some to have been identical with the Dominican Jordanus Saxo, who died at sea in 1237 while returning from the Holy Land; the French Franciscan mathematician Alexandre de Villedieu, who died about 1240; the English mathematician and astronomer Joannes de Sacrobosco (or

John of Halifax), who studied at Oxford but lived most of his life in Paris, where he died about 1250; and the English astronomer and physician William the Englishman (or Marsiliensis), who flourished in Marseilles c. 1231.

Nemorarius made original contributions both to static mechanics, concerning the lever and the inclined plane, and to an idea about a kind of gravity, as well as to the theory of numbers. Villedieu's extremely popular didactic poem on arithmetic, *Carmen de algorismo,* translated into English, French, and Icelandic, greatly helped to spread the use of Hindu numerals throughout Europe. It was the first Latin text that used zero simply as one of the numerals. Sacrobosco's treatises on astronomy (including of course astrology), arithmetic, and the calendar, though lacking in originality, remained enormously popular for several centuries, and his *Algorismus vulgaris,* which referred to the Arabs as the inventors of the algorithm, became the source of the name Arabic for the Hindu numerals.

Though not very accomplished as a practicing scientist, the English Franciscan friar and Aristotelian philosopher Roger Bacon (c. 1220–1292), who puzzled over the problem of the transmission of a mechanical force (a problem that would cause some difficulties for Isaac Newton four centuries later), exerted a great and long-lasting influence upon the later development of science. What singles him out among medieval thinkers is his strongly expressed emphasis on experience and experimentation, in contrast to pure reasoning, in spite of the fact that he also appreciated the value of mathematics in science. Here were the first stirrings of what was needed for a scientific revolution. William of Ockham (c. 1285–1347) was another English Franciscan philosopher who had an impact upon scientific thought. In his case it was the enunciation of what became known as Ockham's razor: "Pluralitas non est ponenda sine necessitate," or "Plurality should not be posited without necessity." Not original but stressed by him with unusual force, this injunction to prefer simplicity over unnecessary elaboration would be taken to heart by many scientists.

Even though the eastern part of what remained of the Roman empire had the intrinsic advantage of speaking Greek, which enabled them to read the pagan natural philosophers and mathematicians without having to wait for translations, the contributions of Byzantine Christian Europe during this time were insignificant. The emperor Justinian had closed Plato's Academy in 529, and the ancient Greek tradition was neglected. Scientific developments outside Christian Europe, on the other hand, were not insignificant. The greatest astronomer of the time was al-Hasan al-Marrakusi, a Muslim living in Morocco, who in the thirteenth century wrote an elaborate treatise on practical astronomy containing a catalogue of 240 stars. This work also included a mathematical part with trigonomical tables of sines, arc sines, and arc cotangents.

At the same time, Islamic Persia produced a great mathematician and scientist, Nasir al-Din al-Tusi (1201–1274), who wrote in both Persian and Arabic and for many years served as astrologer to Hulagu Khan, the Mongol chief who sacked Baghdad in 1258. He wrote works on philosophy, logic, theology, ethics, and poetry, but more important from our point of view, extensive and elaborate commentaries on the works of Euclid, especially on Euclid's fifth postulate, as well as on the works of Archimedes. His principal treatise dealt with plane and spherical trigonometry, the first textbook ever to treat that subject independently of astronomy.

In the service of Hulagu, Nasir al-Din constructed both a library, which contained a large number of volumes that the Mongol armies had "collected" on their march through Persia, Mesopotamia, and Syria, and an observatory in the city of Maragha. He was the first director of this well-equipped observatory, which was said to have several Chinese astronomers and scientists in residence whom Hulagu had brought with him from China. On the basis of his observations there and those of earlier astronomers, Nasir al-Din compiled a large set of astronomical tables listing the motions of planets (ephemerides), as well as astrological data, which were enormously popular as far away as China and remained so for a long time. His treatises on astronomy contained, among other things, detailed criticisms of the Ptolemaic system and ingenious attempts to simplify the complicated machinery of the *Almagest,* though without conspicuous success.

Meanwhile, two widely separated parts of China produced two great mathematicians: Ch'in Chiu-Shao, who flourished from 1244 to 1258 under the Sung on the Yangtze River in the south, and Li Yeh, who lived under the Nu Chen Tartars in the north from 1178 to 1265. The principal work of the first, completed in 1247, consisted of a collection of problems involving equations of higher degree that he solved by a very original numerical method called *t'ien yuan,* or "the celestial element," anticipating much later Western work. (The celestial element was the unknown quantity in an algebraic problem, and its computation included the use of a zero, which may have been imported from India.)[10] Li Yeh, who was primarily an algebraist, remarkably enough also used a slightly different variant of t'ien yuan. If they derived their versions of this procedure from a common origin, this origin was unknown outside the Far East.

A third great mathematician, Chu Shin-Chieh, flourished somewhat later, from 1280 to 1303, but almost nothing is known about his life. The importance of one of his works resides primarily in its influence upon Japanese mathematics after its arrival there via Korea. His second work applied the t'ien yuan for the first time to systems of linear equations with four unknowns. Sarton regards the thirteenth-century development of mathematics in China as "very mysterious."[11]

As the fourteenth century dawned, interest in physical phenomena began to pick up again. Stimulated by a fascination with astrology and consequently with the visually oriented sciences of astronomy and meteorology, interest in optical phenomena reawakened, both in the Christian West and in the Muslim East. St. Thomas Aquinas started it in Western Europe with a commentary on the meteorology of Aristotle, but the common source for this simultaneous study in disparate geographical areas was the *Kitab al-manazir* of Ibn al-Haitham, described earlier. Mentioned both in Genesis and Ezekiel, the rainbow in particular begged to be understood by Biblical scholars.

The leading expert on optical phenomena in the West, the Dominican friar Dietrich von Freiberg (c. 1250–1310), not only carried out numerous optical experiments at a time when experimentation was not yet a common scientific practice but came up with an explanation of the cause of the rainbow that required only relatively minor later modification to be correct. It was produced, he said, by a combination of internal refractions and reflections of light by the spherical water droplets in the atmosphere, a view that was also, quite independently, put forward at about the same time by the Persian Muslim Qutb al-Din al-Shirazi and by his disciple Kamal al-Din al-Farisi. Born in 1236 in Shiraz and living until 1311, Qutb al-Din, who, contrary to Aristotle, regarded light as the source of all motion, made contributions not only to optics but also to medicine and astronomy, in which he introduced important modifications of the Ptolemaic model of the solar system. His student Kamal al-Din contributed an elaborate reinterpretation of the theories of Ibn al-Haitham, including observations on aerial perspective, and he recognized that the speed of light could not be infinite.

Progress in the science of optics led to the invention of spectacles, probably near the end of the thirteenth century, though owing to both technical and psychological difficulties it took a long time before many people would actually use them. The new intellectual ferment in Christian Europe, however, was both broader and deeper than a renewal of the study of optical phenomena, and it was becoming more heated, with the doctrines of the revered Aristotle undergoing modifications in various ways. The thirteenth century was the time of cultural crossover. While the Christian civilization of Europe gained energy, the Muslim-dominated region, for reasons that have never been adequately explained, entered an intellectual and scientific decline.[12] Nevertheless, some Muslim astronomers at that time raised a question that did not occur to any of their contemporary colleagues in Christian Europe—with the singular exception of the Cistercian monk Pierre Ceffons, who carefully demolished all the objections to the thesis but nevertheless refused to accept it: is the earth in motion rather than at rest? On the basis of elaborate proofs, their answer, in every instance, was the same as Ceffons's: it was certainly not moving. Still, raising the question meant taking it seriously.

In fourteenth-century England, the mathematician Thomas Bradwardine (1290–1349) worked in geometry, arithmetic, the theory of proportions, and the Aristotelian laws of mechanics, and died thirty-eight days after being consecrated archbishop of Canterbury. While he stuck to Aristotle's view that the speed with which an object moved was determined by the force acting on it and the resistance it experienced, he reformulated Aristotle's laws of motion so that a doubling of the velocity of an object required the ratio of force to resistance to be squared, rather than doubled, as Aristotle had it. In modern form, this means that Bradwardine postulated $F/R = (\text{const.})^V$ rather than Aristotle's $F/R = \text{const.} \times V$, if F is the force, R the resistance, and V the velocity. The reason for this change was a perceived problem with Aristotle's law of motion when the force is diminished, or the resistance increased, until force and resistance are equal; at this point, any motion was expected to come to a stand-still. But whereas Aristotle's form of the law does not predict this outcome, Bradwardine's does. The old aim of physics to predict the motion of all objects had not been lost sight of.

France produced two philosophers of note in the fourteenth century, Jean Buridan and Nicole Oresme. Buridan is popularly known primarily through a caricature of his writings about human motivation and its determination by desires: "Buridan's ass," standing halfway between two bales of hay and immobilized by desires pulling him in opposite directions, dies of starvation. An Aristotelian philosopher and logician, Buridan (1300–1358) had studied under William of Ockham at the University of Paris and eventually became the rector of that university. He was the first scientific theorist to assume that celestial bodies were subject to the same laws of motion as objects in the sublunar world, a daring conjecture that would be fleshed out and confirmed by Isaac Newton. What is more, Buridan took up Philoponus's concept of inertial "impetus"—in modern language, momentum—which, contrary to Aristotle, kept objects moving even when there was no longer a force to push them. Audacious ideas such as these made him famous all over Europe.

The scholastic philosopher, mathematician, and theologian Nicole Oresme (1320–1382) was of Norman origin, but little is known about his youth and upbringing except that he studied at the College of Navarre at the University of Paris, where he became grand maître in 1356. Serving as a canon at the Sainte Chapelle in Paris and at the Cathedral of Rouen, where he became dean, he was consecrated bishop of Lisieux in 1377. In a development reminiscent of Aristotle himself, Oresme also served as advisor to King Charles V, whose tutor he had been when Charles was the young dauphin. Perhaps because of this association, he adopted the very unusual habit of writing many of his works in French rather than Latin, thereby introducing a number of previously unknown scientific terms into the French language.

An important problem that had been ignored in medieval Europe, though not among Islamic scholars, was now beginning to focus the attention of both Buridan and Oresme: is the earth at rest, with the sun, the moon, and the stars circling about it, or is it rotating, producing in its inhabitants the impression of a diurnal rotation of the heavens? Even to raise this point seriously meant questioning the majestic authority of Aristotle and Ptolemy, as well as that of the Bible. But it had been addressed long ago by Aristarchus and was then more or less forgotten.

Buridan argued that you really could not tell the difference between a sun and body of stars circling the fixed earth and a fixed celestial sphere and sun with a rotating earth: the observed astronomical phenomena would be the same in both cases. Furthermore, if nature preferred simplicity, she would certainly choose to keep the heavens at rest and have the earth rotate. Nevertheless, he could not bring himself to believe that the earth rotated, and his major argument against it was based on his impetus theory: an arrow shot straight up into the air would land west of its launching place if during the time of its rise and fall the earth had moved to the east under it. Since this had never been observed, he concluded from this physical rather than astronomical argument that the earth could not be moving. The same reasoning would be used again and again by others until it was finally put to rest by means of Newton's equations of motion.

Nicole Oresme went on to demolish this conclusion with a counter-argument that placed him well ahead of his time. He decomposed the motion of the arrow into a horizontal and a vertical component. If the earth were rotating, the bow would share the horizontal motion of the ground and impart it to the arrow launched by it. Therefore, its flight would consist of a combination of two motions, a vertical component, straight up, and a horizontal one that would be exactly equal to the horizontal motion of the launcher standing still on the earth. Hence, the arrow would land at its launching point, whether the earth moved or not. In addition, Oresme produced a Biblical argument: as every good Christian knew, God had intervened on the side of Joshua's army, lengthening the day by making the sun stand still over Gibeon. Would it not have been very much simpler for God to achieve the same effect by stopping the rotation of the little earth rather than by bringing both the sun and the entire huge celestial sphere to a halt? This miracle thus could have been performed with greater economy of effort if the earth were rotating. Reasoning based on simplicity, a version of Ockham's razor, was beginning to play a powerful role in scientific thought.

Oresme's ingenious arguments set important precedents for later discussions at the time of Copernicus, but they did not in the end persuade their author, who chose to stick with the Bible, convinced that the earth was at rest. Habits of thought, reinforced by the weight of authority, were difficult to break. His anti-Aristotelian reasoning showed, however, that logic could not prove the immobility of the earth; the basis for that assumption had to be faith, and that conclusion was momentous.

Viewed from the perspective of the pervasiveness of determinism in physics until the twentieth century, the most interesting aspects of Oresme's thinking were his arguments against an old Greek doctrine, resurrected by some neoplatonists and taken up prominently five hundred years later by Friedrich Nietzsche: the eternal return, that every occurrence in history would repeat itself over and over again, ad infinitum. Its astrological underpinnings, of crucial importance at the time, posited that the planets in the sky, which, after all, were believed to determine the entire course of history, could not avoid returning to any position they once occupied and subsequently exactly repeating their earlier motion after a period which the Stoic philosophers called the "Great Year." If the heavenly machinery could not avoid repeating its motion infinitely many times, so necessarily would human history as well, with the length of the "Great Year" estimated to be about 36,000 years.

Oresme's attack on the idea of the eternal return is based on mathematics. In *On the Commensurability or Incommensurability of the Celestial Motions,* he argues that if two objects in regular circular motion have velocities that are commensurable, then there will necessarily have to come a time when they both return to their starting points.[13] Suppose that object A has the velocity V_1 and B has the velocity V_2, and that the ratio of the two velocities is $V_1/V_2 = a/b$, where a and b are two whole numbers. (This is what is now meant by "commensurability"; Oresme's definition was somewhat different, but no matter.) Then the time T for A to complete a rotations is exactly equal to the time for B to complete b rotations, so that at the time T after their start, both will be precisely at their initial positions, and the motion will repeat itself. However, he goes on, if their velocities are incommensurable, so that no such whole numbers a and b exist, then there will never be a time at which the two simultaneously return to their starting point. What is more, he argues (without giving any good reason, except for an analogy with "perfect numbers," which are the sum of their divisors other than themselves—pretty rare) that the probability is much against two planets having commensurable velocities. He then considers more complicated systems consisting of more than two planets in circular motion, and he comes to the same conclusion. If two planets are in conjunction at one time (the astrological event of greatest significance), the probability is against their having been in exact conjunction at any earlier time or being again in conjunction at some future time.

Three things are remarkable about these ideas: first, Nicole Oresme anticipates the nineteenth-century ideas of Henri Poincaré that we will take up later; second, he anticipates more precise mathematical findings about the frequency with which rational numbers occur among the irrational; third, and most important from our point of view, for the first time the notion of probability enters into physical science. In the fourteenth century Oresme's thinking was certainly anomalous, and we shall see nothing of its kind again for five centuries.

As far as other philosophers and natural scientists were concerned, the fourteenth century was still dominated by Aristotle—both the questions he raised and the answers he gave. Since medieval thinking and its educational system were dominated by "questions" and long disputes about each of them, this is meant literally. While men often disagreed about the detailed mechanisms involved in the functioning of the world, the basic "questions" about it were always concerned with whether the world could have existed from eternity or was created, whether a multiplicity of other worlds existed and what that meant, whether the world was "perfect" or corruptible, what material the world was composed of, and how the celestial bodies, the sun, the stars, the moon, and the planets moved. These questions, of course, always skirted along the edges of theological dogma, and thinkers had to beware of offending either authorities or their own Christian beliefs. Observational astronomy, to the extent that it was pursued, primarily served two purposes: the first and foremost, as everywhere in the world, was astrological; the second, however, was the more intellectually constructive one of aiming at a better understanding of how the cosmos worked.

In India, by contrast, where astronomy had long been a prominent and productive pursuit, this science was looked at somewhat differently. Its primary aim was astrological, just as it was in Europe. But it was also considered of divine origin and therefore not subject to changing interpretations; furthermore, its principal non-astrological goal was computational rather than kinematic. Astronomers focused their creativity and energy on making detailed listings and mathematical computations, and, differing from their colleagues in Europe, they did not give much thought to the relation of astronomy to natural philosophy or to physical explanations of the motions of the heavenly bodies.

The combination of the Black Death of 1348 and the Hundred Years' War from the middle of the fourteenth to the middle of the fifteenth century produced a hiatus in the advance of science. Perhaps appropriately, the only significant developments in physics during the fifteenth century concerned natural phenomena that tended to strike fear into people: several impressive comets were observed during that period. The dominant explanation for the nature of comets that Europe inherited from Greek antiquity was provided by Aristotle, who did not regard them as celestial objects because he defined the latter as "eternal and not subject to increase or diminution" (*On the Heavens* 1.3).[14] He disagreed with Anaxagoras and Democritus, who thought of comets as conjunctions of known planets, and with the Pythagoreans and others, who believed comets to be planets that became visible only occasionally, the tail caused by a reflection of the sun on moisture attracted to the planet. Instead, Aristotle's view was that comets were localized fires caused by friction between the upper layer of the atmosphere, the terrestrial sphere, which is constantly in irregular motion, and the steadily moving celestial sphere. They were therefore located at the edge of the upper atmosphere, relatively close by, and their existence was a simple consequence of his cosmology. Regarding them entirely as astrological phenomena, Ptolemy accepted this Aristotelian view of the proximity of comets and mentioned them almost nowhere in his writings.

Such power did the thought of Aristotle (supported by Ptolemy) exert that until the fifteenth century neither Christian Europe nor the Islamic world generated new ideas about comets or even paid much attention to them. However, the appearance of one during most of the months of February and March in 1402 occasioned a lengthy treatise by Jacobus (Engelhart) Angelus of Ulm, probably a native of Swabia, who served as physician to Prince Leopold, duke of Austria. The treatise describes detailed observations of the comet at various times and discusses its astrological significance. The later comets of 1433, 1449, 1456, 1457, and 1472 were observed and described with unprecedented precision and detail by the Florentine Paolo Toscanelli dal Pozzo, who lived from 1397 to 1482. Acquainted with Brunelleschi, who was then in the process of constructing the basilica of Santa Maria del Fiore in Florence, with its magnificent dome, Toscanelli installed a gnomon—the highest ever built—in the cupola of that church and used it for precise astronomical observations such as the date of the summer solstice. However, neither Angelus nor Toscanelli had anything new to add about the nature or distance of comets.

The first astronomer to take a stab at estimating the distance of a comet from actual parallax observations was Georg Peurbach, who was born 1423 in Peurbach, Austria, and died in 1461 in Vienna, where he had taught at the university. Parallax is the altered angle at which an object is seen when viewed from different positions. That angular difference makes it possible to determine the distance between the object and the observer. He too left detailed descriptions of the 1456 comet, and on the basis of his and Toscanelli's precise observations it could later be identified as the same comet that returned in 1758. Since this comet was predicted by the astronomer Edmund Halley in 1705, it became known as Halley's comet. Astrological myth has Halley's comet visible at both Mark Twain's birth in 1835 and his death in 1910. Though the visibility of the comet is apocryphal, the years are correct. Taking Aristotle's assumption that comets are located at the interface between the terrestrial and the celestial spheres more or less for granted, Peurbach attempted to estimate the height of that region by a parallax measurement (Fig. 8). However, as his figures were mistaken and his calculations in error, he arrived at results that were quite incorrect.[15]

Johannes Müller, born in 1436 in Königsberg, Bavaria, who became widely known under the name of Regiomontanus (a latinization of the name of his birthplace), studied at the University of Vienna under Peurbach, receiving his bachelor's degree there at the age of fifteen. Owing to university regulations, he had to wait until he was twenty-one to receive his master's degree and become a member of the faculty in Vienna. Considered the greatest astronomer of the fifteenth century, he spent much time and effort on attempts to determine the size and distance of comets, which he regarded as having no proper motion. This erroneous assumption may have been the fundamental source of his faulty parallax observations, and he never discovered that comets are in fact much more distant than Aristotle had taught; that discovery had to wait for Tycho Brahe in the sixteenth century.

Perhaps Regiomontanus's most fruitful work in the long run was his critical annotated Latin translation, in cooperation with his teacher Peurbach, of Ptolemy's *Syntaxis*, under the title *Epitome*, in which he criticized Ptolemy's model for variations in the apparent lunar diameter that were greatly at odds with observations. Copernicus's reading of this critical passage was one of the stimuli for producing his

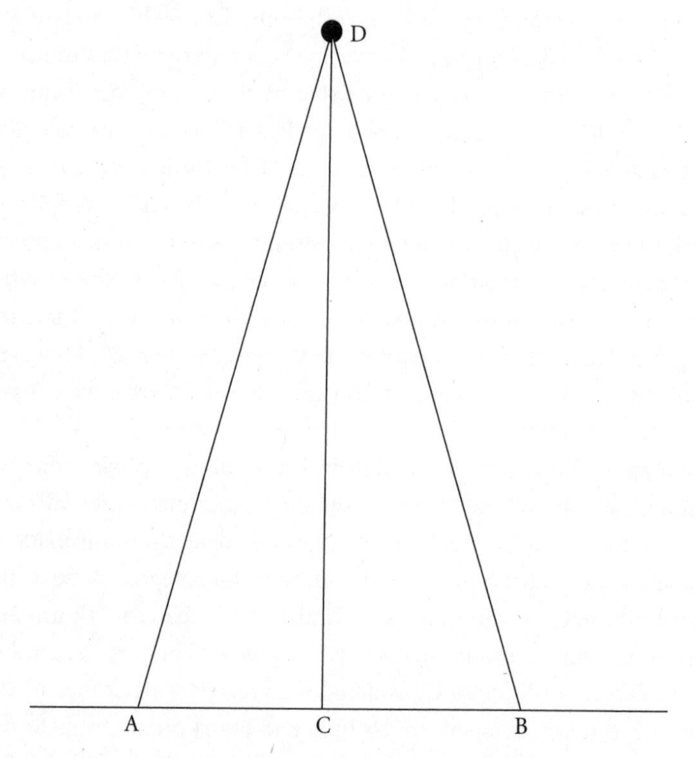

Figure 8 The ratio of the distance of the object D from the point C to the distance from A to B can be determined from the difference in the direction of AD from that of BD (the parallax).

revolutionary new model of the solar system. Regiomontanus died in Rome in 1476, probably of the plague, but rumors circulated that he was poisoned by a professional rival.

At the threshold of the sixteenth century, two more mathematicians of some importance made an appearance. The first, Niccolò Fontana, son of a poor postman, was born in 1499 in Brescia, near Venice. He had been severely injured when French troops sacked his home town and was left with a permanent speech defect that gave rise to his nickname, Tartaglia, the stutterer. Tartaglia grew up largely self-educated, particularly in mathematics and physics, reading whatever sources he could find. Eventually he became professor of mathematics in Venice, and he died there in 1557 in very humble circumstances. His work toward the solution of cubic equations was his principal mathematical contribution; in physics he applied his understanding of mechanical problems primarily to matters of military defense. He was the first to state what is still sometimes known as Tartaglia's theorem: the trajectory of a projectile is a curved path, with the maximum range attained if the firing elevation is 45° (in vacuum), though he provided a faulty proof for this correct proposition. Tartaglia also translated both Euclid and Archimedes into Italian, their first rendering in a modern language.

The second was Girolamo Cardano, born in 1501. Though he was widely known at the time primarily as a physician and was the first person to give a clinical description of typhus fever, his book *Ars Magna* played an important role in the history of algebra. (Note here the first mention of a book: Gutenberg's Western invention of the printing press employing movable type had taken place in the middle of the fifteenth century; his Bible was first printed in 1455.) It contained the solution of the cubic equation, based on the work of Tartaglia, as well as a solution of the equation of fourth order, which he owed to his former servant Lodovico Ferrari. Cordano's *Liber de ludo aleae* was the first book dealing with the systematic computation of probabilities.

Inconsolable over the loss of his favorite son, who was executed in 1560 for poisoning his wife, Cordano was arrested for heresy in 1570 and never again permitted to publish a book. He died in 1576 in Rome. In reaction to the Reformation, the Catholic Church by this time was cracking down on any perceived enemies.

four

The First Revolution

The flowering of science in Western Europe that began in the sixteenth century did not sprout in a desert. The medieval period had already seen the great works of Dante Alighieri and Chaucer, and the Renaissance would soon produce Cervantes and Shakespeare. The Italian rebirth in architecture and the arts was in full swing, begun, by the usual reckoning, with the painter Giotto at the beginning of the fourteenth century, followed by Brunelleschi, Leonardo da Vinci, Michelangelo, and Raphael, to mention but a few of the greatest. As far as our image of the universe is concerned, the opening shot of the first scientific revolution was reluctantly but decisively fired by Nicolaus Copernicus, and it reverberated far and wide.

Copernicus was born in 1473 in the city of Torun on the Vistula River in Poland, south of the Baltic seaport Gdansk. The son of a well-to-do merchant who died when the boy was ten years of age, Copernicus grew up in the house of his uncle Lucas Watzenrode, soon to be bishop of Varmia. He began his studies in liberal arts at the University of Krakow and continued in Bologna, where the astronomer Novara introduced him to two works that turned out to have a seminal influence on his thinking: Pico della Mirandola's *Disputations against Divinatory Astrology* and Regiomontanus's critical translation *Epitome of Ptolemy's Almagest.* (Astronomy and astrology were at that time regarded as two subdivisions of "the science of the stars.") The first pointed out that since astronomers had never agreed among themselves about the order of the planets, astrologers were foolish to base their divinations on the different powers exerted by various planets, as these powers depended on unreliable data about their order. The second found glaring discrepancies between Ptolemy's geocentric model of the solar system and actual observations. The seeds of doubt concerning accepted views of the universe, based on both Aristotle and Ptolemy, were thus planted in the young man's mind.

After continuing his studies at the University of Padua, primarily in medicine (a subject that at the time was intimately connected with astrology), Copernicus received his doctorate in canon law at Ferrara, obtaining a sinecure at Wroclaw, which his uncle had arranged for him, and a canonship in the cathedral chapter of Frombork (Frauenburg). As these positions left him with plenty of spare time for his primary interest, astronomy, he built himself a little roofless stone tower in Frombork that he used as an observatory (though he never actually made many observations of his own). He died in Frombork in 1543.

In his first publication, called *Commentariolus,* Copernicus challenged the views of Aristotle and Ptolemy by proposing a heliocentric planetary system, without claiming any priority for this idea. His model for the solar system, with the sun at the center and both the earth and the planets revolving in circles about it, allowed the planets to be arranged in an orderly and unambiguous way, with increasing orbital periods: Mercury (88 days), Venus (225 days), Earth (1 year), Mars (1.9 years), Jupiter (12 years), and Saturn (30 years). His later work, *De revolutionibus orbium celestium libri vi,* laid out this system in all its detail, but he was so reluctant to publish it that he delayed its printing for 36 years, until a year before his death. He also took great care not to mention Aristarchus, because he knew that this earliest proponent of a heliocentric system had suffered for the audacity of his unorthodox proposal by being accused of impiety.

In addition to having the sun at the center of the universe and having the planets, including the earth, revolve about it, the Copernican solar system had the earth spinning about an axis through its center that tilts away from the vertical with respect to the plane (the ecliptic) of the earth's orbit about the sun. Moreover, that tilting axis itself slowly wobbled, describing a cone, like the axis of a spinning top, thereby accounting for the precession of the equinoxes. The universe, he argued, had to be very large because otherwise the orbital motion of the earth would give the stars a noticeable parallax, which had never been observed. On the other hand, his universe was not infinite, because if it were, it would make no sense to place the sun at its center. (Where is the center of an infinite space?) In contrast to all earlier theories, which he portrayed as disorderly monstrosities, he depicted his newly created image of the world as an orderly, harmonious composition, no part of which could be altered without destroying the whole.

What made the ideas of Copernicus revolutionary was not simply the suggestion of a heliocentric rather than a geocentric solar system; such proposals had been made, and forgotten, several times in antiquity. It was the fact that he insisted on presenting his system not merely as a model, convenient for astronomical computations, but as physically real: he meant it to be a true representation of the cosmos. Many of the readers of *De revolutionibus* nevertheless accepted his proposal as no more than a model because that was easier than coming to terms with its revolutionary implications. Such a view of reality implied a complete break with Aristotle, who had made a fundamental distinction between the sublunar world and the sphere beyond, the first filled with earth, fire, water, and air and the second with ether. Moreover, objects in these two regions moved according to basically different laws. This picture of the universe was incompatible with what Copernicus proposed and would have to be abandoned to accommodate his views.

Not to mention the contradiction between the Copernican proposal and the Bible. His system shrank human beings from their primary place in God's creation at the center of a relatively small, homelike cosmos to creatures inhabiting a minor planet eternally revolving about the sun, all located in a vast space occupied by the stars. Very intricate mechanical clocks, still rare and unreliable but much admired, existed by that time, and the young Georg Rheticus, an ardent Lutheran follower of Copernicus, compared this newly proposed universe to the well-oiled wheel mechanism of a clock. A flag-waving barricade-stormer Copernicus was not, but his message was revolutionary all the same.

And the Church recognized it. The first Copernican to bear the brunt of Rome's wrath, though largely for unrelated reasons, was the Dominican priest Giordano Bruno. The son of a professional soldier, Bruno was born in 1548, five years after the death of Copernicus. Following his studies of the humanities in Naples, he was attracted to the philosophy of Averroes as well as to Arianism, a heretical doctrine that denied the divinity of Christ. Bruno came to believe that he was reviving the religion of ancient Egypt, with magical insights into nature that enabled him to understand the Copernican system better than Copernicus himself. He envisaged an infinite universe that contained innumerable animated worlds. Fleeing from Italy to Geneva, he became temporarily a Calvinist, and then moved on to Paris and London. Always provocative and repeatedly excommunicated by various churches, he stayed at the court of Queen Elizabeth for some years, attacking the old-fashioned Aristotelians at Oxford, where he lectured on the Copernican system.

Finally, Bruno made the fatal mistake of returning to Italy, where he was denounced to the Venetian Inquisition for heresy. After saving himself by partially recanting his views there, he was extradited to the fiercer Roman Inquisition. During a trial in Rome that lasted eight years, he refused to retract anything and was finally convicted. In 1600 he was burned alive in the Campo de' Fiori.

The fiery death of Giordano Bruno has sometimes been misinterpreted as the iconic expression of the Catholic Church's reaction to the scientific revolution begun by Nicolaus Copernicus. Occasionally, Bruno's upright refusal to retract in Rome has been favorably compared with Galileo's "craven" recanting at his later trial there. However, Bruno's scientific views and his advocacy of the Copernican solar system were not the real cause of his condemnation by the Inquisition. He was condemned by the Church for heresy because of alleged diabolical magical practices and because he denied the divinity of Christ. Nevertheless, the fate of this Copernican would serve as a warning for a long time to come: scientists who angered the Church were disturbing a dragon.

The Danish astronomer Tycho Brahe, an almost exact contemporary of Bruno, does not really belong in the company of the revolutionaries. Nonetheless, his work paved the way for Kepler, thus enabling the revolution to continue. Tycho Brahe, whose father was governor of the castle of Helsingborg, was born in 1546, abducted at a young age by his wealthy uncle, and raised in the latter's castle at Tostrup, Scania. He began his education by studying law at the University of Copenhagen and continued, three years later, at Leipzig. However, the experience of witnessing a total eclipse of the sun in 1560, together with the fact that the date of the startling event had been predicted with near accuracy by astronomers, made such an impression on the fourteen-year-old boy that he decided to devote his life to observing the stars.

Listening to lectures on law in the daytime, he studied Ptolemy's *Almagest* in the evening, and with the help of some of his teachers he began to construct instruments for measuring the positions of stars and made his first small globes for plotting them. At the age of seventeen he observed and recorded his first planetary conjunction, between Jupiter and Saturn, and discovered that all the existing astronomical tables, including those of Copernicus, were grossly inaccurate in their predictions of this event. He resolved then and there to change this state of affairs and to spend all his efforts making and recording accurate observations of the heavens and correcting all the existing tables.

After losing part of his nose in a duel at the age of twenty and having it replaced by a cover made of gold, silver, and copper, Tycho traveled throughout Europe, acquiring a large number of astronomical instruments on the way. Upon his return he settled on his inherited estate and built himself a small observatory. On November 11, 1572, he suddenly saw a new star in the constellation of Cassiopeia, brighter than Venus, which, by careful observation, he ascertained to be beyond the distance of the moon. Disturbing the perfect and unchanging Aristotelian harmony of the celestial sphere, this first observation of a nova in Europe (several had already been recorded in China, but Europeans were unaware of this) was a very unsettling phenomenon, both to him and to everyone else. Its publication instantly made the young Tycho famous all over Europe. In order to keep the celebrated Dane in Denmark, King Frederick II granted him title to the island of Ven in the middle of the sound near Copenhagen, as well as financial support for research and for the construction of an observatory, which he called Uraniborg, after Urania, the muse of astronomy. There, he and his staff made and recorded with unprecedented accuracy many important observations that substantially corrected nearly all the known astronomical tables, just as he had resolved to do at age seventeen.

When his friend the king died, Tycho lost most of his income and, under the patronage of the Holy Roman Emperor Rudolph II, he moved to Prague. Four years later, in 1601, he died there, leaving all his accumulated data to Johannes Kepler, who had been his assistant during his final year. His very precise observations—remember, this was before the invention of the telescope—had shown him the inadequacy of the Ptolemaic system and how much of an improvement, though still imperfect, the Copernican model was. But he still had not been able to bring himself to accept the idea of the sun at the center as a reality, and in a step backward from Copernicus, he had devised his own configuration, in which the sun orbited a stationary earth and the planets revolved about the sun. Tycho's data would soon serve to bolster the next stage in the revolution, Kepler's new model of the solar system.

Johannes Kepler was born in 1571 in the small town of Weil der Stadt near Stuttgart in Germany, son of a mercenary soldier who abandoned his family. A small, frail man, he was bothered by stomach ailments and fevers all his life. Though he did not like to rock the boat—as he later declared in the dedication of one of his books, "I like to be on the side of the majority"—that is just what he turned out to do, with great effect.

A ducal scholarship enabled him to attend a Lutheran seminary at the University of Tübingen with the intent of becoming a theologian, though his interest in astronomy and astrology had already been aroused by his early viewing of the comet of 1577. At Tübingen, the astronomer Michael Mästlin introduced him to the works of Copernicus, of whom Mästlin was an ardent but publicly cautious admirer, tutoring him in the details of the system. For Kepler, the Copernican universe showed the unmistakable mark of divine planning, including the creation of the human mind to understand and celebrate the structure of the world. The solar system, as presented by Copernicus, represented for Kepler a pure symbolic image of the Trinity, and he would give up theology and spend his life substantiating all its glorious perfection.

After teaching mathematics and astronomy in Graz for several years, where he published his first major paper and fell into some trouble as a Lutheran, Kepler moved to Prague in 1600, becoming Tycho Brahe's assistant a few months before the great astronomer's sudden death, and subsequently his successor as imperial mathematician to emperor Rudolph II. In 1610 he was given his first telescope—just invented in Holland and improved by Galileo—and after the deposition and death of Rudolph he moved to Linz as district mathematician for Upper Austria, later becoming the private mathematician of General Wallenstein, duke of Friedland. For a number of years, Kepler was plagued by personal problems—his mother was accused of witchcraft but finally exonerated, his wife died, one of his sons succumbed to smallpox—and he was continually caught up in the religious strife of the time, but he was a remarkably resilient man.

His first step in the direction of showing the splendor of the cosmos as conceived by Copernicus occurred to Kepler in 1595, when he remembered Plato's five perfect polyhedra. Striving to satisfy his strong sense of order and harmony, he thought he had found the key to the perfection of the universe. His calculations showed that if he placed the orbits of Copernicus's planets between the five regular solids, nested as Plato had suggested, their radii would come out correctly to within 5 percent—except for Jupiter. But that planet was so far away that it was no wonder it was exceptional. The long-lasting influence of Plato's mode of thinking was not conducive to scientific progress, and misguided attempts to explain the relative distances of the planets from the sun would continue to be made occasionally without success as late as the nineteenth century.

Kepler, whose ruling belief was that the functioning of the solar system had to be explained by physics, clarified his purpose in a letter to a friend with these words: "My aim . . . is to show that the celestial machine is to be likened not to a divine organism, but rather to a clockwork . . . in so far as nearly all the manifold movements are carried out by means of a single, quite simple . . . force, as in the case of a clockwork [all motions are caused] by a simple weight."[1] His inspiration for a fresh way of looking at the planetary system was the work of William Gilbert (1544–1603), the most distinguished scientist at the court of Queen Elizabeth, who had explained the workings of the compass by introducing the idea of an attraction reaching from one magnet to another. Kepler concluded that the motion of the planets must be somehow centrally directed by the sun. Following Gilbert's theory, he therefore thought magnetism must be the causative agent, the weight that made the clock run. Specifically, what kept the solar system in motion was that the earth and the planets were magnets dragged around by the rotating sun through its magnetic force, the strength of which decreased with the distance. A minute examination of the high-precision data he had inherited from Tycho Brahe convinced him that Mars could not be moving uniformly in a circle, and the attenuating pull from the sun could easily account for its slowing down at greater distance if the orbit were not circular.

After a temporary diversion in 1604—the discovery of another new star, later called Kepler's Nova (now known to have been a supernova)—he announced three novel rules of planetary motion; these are nowadays usually referred to as Kepler's three laws. (Not published all at once, the three rules appeared in various of his publications.) The first law states that the orbit of each planet, including the earth, is not a circle but an ellipse, with the sun at one of its foci. Destroying the beautiful perfection of the circular orbits that in one way or another—in Ptolemy's model by piling epicycles upon cycles—had dominated all previous models of the solar system, this feature was regarded by many as unacceptably ugly. The second law says that the straight line from the sun to a given planet traces out equal areas in equal times, implying that when a planet is traversing a part of its orbit that is closer to the sun, it moves faster than when farther away. Again, the comforting beauty of uniform motion in the heavens was abandoned. The third law states that the squares of the periods of the planets are proportional to the cubes of their mean distance from the sun, a rule that especially satisfied Kepler's mathematical mysticism and made the whole system quite harmonious in his eyes. Finally, building on his predecessor's earlier labors, he completed Tycho's precise planetary and stellar records, known as the

Rudolphine Tables, which became invaluable for later astronomers as well as for astrologers.

In addition to his path-breaking astronomical work, Kepler also contributed to the physics of optics, particularly concerning the phenomenon of refraction and its effect on vision. Though eyeglasses had been invented more than two hundred years earlier, he was the first to explain why they actually worked, and he wrote a treatise on the optics of the telescope that included a description of a novel type with two convex lenses. He also dabbled in the theory of tiling, coming up with some quite remarkable designs.[2] Curiously, Kepler wrote a book of rather prescient science fiction, *Somnium seu astronomia lunari,* that describes a dream-fantasy voyage to the moon, whose overtones of witchcraft ended up playing an embarrassing role at his mother's trial. After General Wallenstein, his patron at the time, lost his position as commander-in-chief in 1630, Kepler traveled to Regensburg, where he became seriously ill and died that same year. His grave has been lost, a victim of the Thirty Years' War.

The Islamic world during this period was dominated by the Ottoman Empire, which had arisen in the fourteenth century and conquered Constantinople in 1453. Islam bore no religious preconceptions concerning the earth as the center of the universe. Since its civilization had a strong tradition in astronomy, and some of its models of the solar system could be regarded as mathematically equivalent to that of Copernicus, the scientific developments in Christian Europe were naturally of great interest to Ottoman astronomers. This was especially so because some of the new, more accurate stellar tables generated in the West might be relevant to the lunar-based Muslim calendar. The new tables were therefore quickly imported, and Sultan Selim III ordered fresh calendars to be published in conformity with these improved data. The controversy over the reality of the heliocentric versus the geocentric solar system that generated so much heat in Christian Europe left Muslims cold. They looked at the change from one system to the other simply as a change of coordinates, a technical matter of no religious or metaphysical significance.

The work of Copernicus and Kepler effectively freed European astronomy from the powerful influence of Aristotle and pointed in the direction of explaining the functioning of the heavens by the same physical laws that governed the motions of objects on earth, transforming the regular celestial movements into a clocklike machinery. These underlying physical laws, however, were still thought to be the same as those promulgated by Aristotle. It remained for Galileo and Newton to sever the last strings that bound the ancient philosopher to accepted physical science.

Born in Pisa in 1564, the same year as Shakespeare, Galileo Galilei, the eldest of seven children of a musician, was educated at the monastery of Santa Maria at Vallombrosa near Florence and at the University of Pisa as a medical student. A lively man with a pugnacious disposition, full of sarcastic wit but rarely personally disliked, he was respectful of authority in matters of religion and politics but could be offensive to administrators and scathing to his philosophical adversaries.

When the subject of medicine failed to hold much interest for him, he began to take private lessons in Aristotelian physics and in mathematics, particularly devoting himself to Euclid and Archimedes. He also made his first significant discovery: the isochronism of the pendulum. He found that the period of a simple oscillating pendulum, such as a chandelier swinging in the wind, did not vary with the width of its swing (so long as it did not swing too widely). This discovery would turn out to have a profound influence on the construction of reliable clocks and on the most basic fields of physics over the next four centuries.[3]

After delivering invited lectures on a mathematical treatment of the geography of Dante's *Inferno* and applying unsuccessfully for the vacant chair in mathematics at the University of Bologna, he was appointed to the mathematics chair at Pisa. At this time his father was engaged in controversies over the connection between the length of strings of musical instruments and the resulting harmonies, which first stimulated Galileo's life-long emphasis on testing mathematical relations among physical phenomena by detailed observations. Even though for him such observations more often served as convincing demonstrations than as experimental searches for new knowledge, this approach represented a definite break with the previous practice of relying entirely on the power of reasoning to arrive at physical truth.

As time went on, Galileo became completely disenchanted with the prevailing Aristotelian legacy in physics, and he made no bones about it to his philosophical colleagues. The famous spectacle of his dropping cannon balls of different sizes from the leaning tower of Pisa probably never took place, but if it did, it was surely meant as a public exhibition to the philosophers that Aristotle had been wrong to claim heavier objects fall faster than lighter ones, rather than as an experiment to find out if the ancient philosopher was wrong or right. His first work on the motion of bodies was a treatise, commonly referred to as *De motu,* which dealt not only with falling objects but also with their speeds of descent on inclined planes. As his initial conclusions here were contradicted by experiments, he was led to the recognition that, contrary to Aristotelian teachings, it was acceleration rather than just speed that played the essential role in the laws governing moving objects, a point that would be crucial in his understanding of movements along circular paths. Even though his new perspective amounted to a complete denial of Aristotle's split between "natural" and forced motion, he still stuck with a geocentric view when applying his ideas to the earth. But he was at that time simply not yet very interested in astronomy.

When friction with his colleagues and with the university administration (he ruffled feathers by, among other things, writing a poem poking fun at the wearing of academic robes) led to the loss of his position at Pisa, Galileo was appointed to the chair of mathematics at the University of Padua. At this gathering place for scholars from all over the continent, the intellectual atmosphere was more conducive to his further development. His notes on mechanics, written for his students, began to circulate throughout Europe, translated into both French and English. By this time, he had become convinced of the virtues of the Copernican system, probably more because of its ability to use the motion of the earth to account for physical effects like the tides than because of astronomical arguments. Galileo's view that the tides could not be explained without taking the motion of the earth into account was correct, though his actual theory to account for the tides was not.

When he received a copy of Kepler's first book, Galileo wrote to its author that he supported Copernicanism, but he did not do so openly. In any event, he did not like Kepler's elliptical planetary orbits, as their elongated shapes offended his aesthetic sense. Pursuing with renewed interest the problem of accelerated motion, both in free fall and along circular arcs, he began to write a systematic treatise on the subject. Also while in Padua, he took a Venetian mistress who bore him two daughters and a son. When he left Padua, she remained in Venice and eventually married another man.

The supernova of 1604, Kepler's Nova, which made a considerable stir in Europe, finally awakened Galileo's interest in astronomy. But the crucial event for him occurred a few years later, when he received word of the invention of the telescope by the Dutch lens grinder Hans Lipperhey. Galileo immediately set to work constructing such a device for himself, much improving it in the process, so that by 1609 he had achieved about thirty-fold magnification. The Venetian government, which as a maritime power immediately recognized the value of a telescope for naval purposes, offered him a lifetime appointment to its university at an unheard-of salary, but he declined and returned to Florence to accept a position as philosopher and mathematician at the court of the grand duke of Tuscany and as professor of mathematics at the University of Pisa, without any teaching duties.

In 1610 Galileo turned his telescope to the heavens—he was the first one to use the device in that way—and what he saw startled him: the moon was full of mountains, the Milky Way was a collection of a vast number of stars, and Jupiter was accompanied by four satellites of its own. Within three months he managed to write it all up, extremely accurately, and to publish it with the title *Siderius nuncius*, or *The Starry Messenger.* The book created excitement throughout Europe, as well as considerable controversy, which the author uncharacteristically ignored; he was too busy observing the sky and making new discoveries. Next were the rings of Saturn, which he mistook for satellites because his telescope was too weak to resolve them, and, more important, the phases of Venus, which added confirmation for the Copernican system. When he traveled to Rome to exhibit his telescope and talk about his discoveries, he was honored by the Jesuits, several cardinals, and the pope himself, and was made a member of the Accademia dei Lincei, the first scientific society in existence.

After publishing a treatise about the behavior of objects under water, favoring Archimedes over Aristotle, Galileo became involved in a controversy concerning sunspots. This led him to write a book in which he introduced the important concept of conservation of angular momentum. The angular momentum of an object rotating about a central point is equal to its momentum multiplied by its distance from that center. For the first time, Galileo supported the Copernican system in print. During the disputes that arose as a result, he took the position that theological interference in purely scientific questions was inadmissible. Neither the Bible nor nature, he asserted, could speak falsely; however, nature was the province of scientists, and it was up to theologians to reconcile the facts discovered by scientists with the language of the Bible. Needless to say, this did not endear him to the Church.

Against the advice of his friends, he traveled to Rome in 1615 to clear his name. He personally succeeded in this, but Pope Paul V also set up a commission to investigate the theological status of the motion of the earth. On the basis of its conclusions, Galileo was officially instructed in 1616 to refrain from either holding the Copernican view or defending it. Returning to Florence, he busied himself with less controversial problems, such as using the eclipses of his newly discovered satellites of Jupiter as a universally visible celestial clock that could be used for the determination of longitude at sea. This proposal turned out to be impractical, as Jupiter's satellites are too difficult to see from a wind-tossed ship.

While Galileo was engaged in research concerning comets—three of them had startled Europe in 1618—and in further controversies as well, an old friend and patron of science, Cardinal Maffeo Barberini, became Pope Urban VIII. To pay his respects, Galileo journeyed to Rome in 1624 and obtained the pope's permission to write a book discussing the Copernican system, provided he impartially presented both the Copernican and the Ptolemaic view. The *Dialogue Concerning the Two Chief World Systems* took him six years to write, and it was hardly impartial. He first tried to have it published under the auspices of the Accademia dei Lincei, but Rome was tardy in giving permission. When he quickly managed nevertheless to have it published in Florence in 1632, it initially caused no problems. After a short delay, however, the printer was ordered to cease further sales and Galileo was summoned to Rome.

A hostile ecclesiastical faction at the Vatican had been able to persuade the originally quite friendly Pope Urban that Galileo had both caricatured him personally in the *Dialogue,* putting Barberini's arguments in the mouth of a simple-minded Aristotelian, and deceived him by concealing that he had been under a personal injunction from Pope Paul never to discuss the Copernican system. (This last point was either incorrect or based on a misunderstanding; Galileo felt himself to be innocent of the charge of deception.) As a result, the case against him was vindictively pursued with great force, and the pope insisted that he come to Rome to face trial, even though it was the middle of winter and Galileo was seriously ill; otherwise he would be brought there in chains. Finally, the grand duke, having exhausted all his own attempts to help, arranged for him to be taken as comfortably as possible by litter to Rome in 1633.

At the trial, during which the memory of Giordano Bruno's fate could not have been far from his mind, Galileo was made to abjure the "heresy" of Copernicus, the *Dialogue* was placed on the index of forbidden books, and he was sentenced to life imprisonment. (The judgment of heresy against the Copernican system and Galileo's discoveries that supported it was not officially rescinded until 1992, by Pope John Paul II. Galileo's trial itself was not declared in error, however, because in contrast to Giordano Bruno, Galileo ran afoul of the Church fundamentally not because of heresy but because of insubordination.) The prison sentence was immediately commuted to supervised house arrest, and he was sent to Siena, under the charge of Archbishop Piccolomini. Treating him with great consideration, the archbishop—possibly a former student of Galileo's—managed, within a few weeks, to raise his crushed spirits, and he began to make plans for writing up all of his results in physics.

In order to avoid another scandal when the Roman Inquisition, unavoidably, got wind of the honored treatment of the scientist by the archbishop in Siena, Galileo was finally moved in 1634 to his house at Arcetri near Florence. As he was leaving Siena, he probably uttered the famous words "Eppur si muove" ("And yet, it does move") which legend has him muttering under his breath on his knees after abjuring the Copernican heliocentric system with its moving earth.

Galileo had been particularly anxious to get back to Arcetri, to be near his oldest daughter, Virginia, who lived in a convent under the name of Maria Celeste and who had been extremely helpful to him during the time of his dispiriting experiences with the Inquisition. Shortly after his return, to his great sorrow, Virginia died after a brief illness, and it took him some time to recover from the resulting depression. Recover, however, he did, and he finished his final work, *Discourses and Mathematical Demonstrations Concerning Two New Sciences,* usually referred to simply as *Two New Sciences,* within less than a year. To have it printed, though, presented another obstacle: the printing of any of his books, old or new, had been forbidden by the Congregation of the Index. So a manuscript copy was surreptitiously taken to France and thence to the Netherlands and finally to Leiden, where it was printed in 1638. By that time Galileo was 74 years old and completely blind.

The sciences referred to in the title of *Two New Sciences* are those of the strength of materials, essentially an engineering subject, and the mathematical science of motion. Galileo discusses a large number of problems relating to the constitution of matter, the nature of sound, the speed of light, and the weight of air, as well as the nature of mathematics and the place of reason and experiment in physics, in addition to detailed treatments of uniform and accelerated rectilinear motion and the parabolic trajectories of projectiles. Objects falling freely, or descending on an inclined plane, move with uniform acceleration, though they do so only in a vacuum, he states; in a resisting medium such as air, they attain a fixed terminal velocity. He also devotes considerable attention to the swinging motion of the pendulum, whose isochronism was his earliest discovery, giving the relation between its period and its length and pointing out that this period did not depend on the mass of its bob. However, he did not manage to explain the reason for its isochronism.

In his mathematical discussions he distinguishes between finite, infinitesimally small, and infinitely large quantities, without flinching from paradoxes that arise and appear unresolvable. On the subject of infinite quantities, for example, he points out that such concepts as "less than," "greater than," and "equal to" are not necessarily applicable, which he exemplifies by showing that the infinite set of natural numbers can be put into one-to-one correspondence with the set consisting of their squares. These are problems with a definitely modern flavor that would preoccupy mathematicians in the nineteenth century.

After the publication of his last book, the great scientist lived for another four years. His son Vincencio took notes on his reflections and assisted him in the design of a clock employing the pendulum and an especially designed escapement. On orders of Grand Duke Ferdinand II, a large clock based on Galileo's plan was eventually built and installed twenty-five years after his death in the tower of the Palazzo Vecchio in Florence, where it still remains, accurate to within one minute per week. He died at Arcetri in 1642 and was buried at Santa Croce in Florence. Contrary to the wishes of the grand duke, no suitable tomb could be erected for fear of offending the Holy Office—Pope Urban had even denied Galileo's request to attend mass on Easter and to consult doctors in Florence about his failing eyesight—until almost a century later, when his grave was marked with an appropriate monument and inscription.

The primary characteristics of Galileo's approach to physics were, first, his reliance on observational or experimental evidence rather than pure reasoning to demonstrate the truth of a statement about nature. The world we see was, in his opinion, real, not an imperfect image of an ideal Platonic universe that could be constructed by the human mind. Second, he emphasized the mathematical description of observed physical processes. The language of nature, he believed, was mathematics, and it was impossible to understand the natural world without knowing that language. His jealously guarded independence of judgment on matters subject to his own direct observation against the pressures of traditional authority exerted a strong general influence on the educated public all over Europe, as did his support of Copernican astronomy. And although his immediate scientific influence was not very great, in the long run his mode of thinking penetrated deeply into subsequent physics.

That he felt no need to search for efficient causes of events when a mathematical description was available distinguished him from many of his immediate successors, but this mode of thinking would eventually become dominant and remains fashionable to this day. Galileo's formulation of physical laws in mathematical terms began to transform the idea of a clockwork solar system into a more general view of nature as a whole, run, like a machine, by gears consisting of algorithms (in modern terms, like a computer). Carried further by the influential René Descartes and by Christiaan Huygens, this transformation led to the work of Isaac Newton, which was the culmination of the first scientific revolution.

Descartes was born in 1596 in La Haye, Touraine. Following eight years at a Jesuit college, studying grammar, literature, philosophy, theology, and his favorite subject, mathematics, he entered the University of Poitiers to study law. Upon graduation, he joined the army of Prince Maurice of Nassau as a military engineer, but decided after a vivid dream that he would devote his talents to the goal of interconnecting all the sciences with mathematics and reducing all of physics to geometry. He began his efforts with algebra. Six years later he returned to France, sold his estate, and spent seven years traveling and meeting scientists throughout Western Europe. For the next twenty years he settled in the Netherlands. At the age of 53, Descartes accepted an invitation to instruct Queen Christina in Stockholm, where he became ill and died in 1650. His remains were returned to France and buried in Paris.

Descartes' greatest contribution to mathematics, which turned out to be enormously fruitful in its application to physics as well as other areas of science, was the introduction of a way of reducing all of geometry to manipulations of numbers. It was the beginning of the field of analytic geometry. With the help of what are now called Cartesian coordinates, this technique allows geometrical problems to be solved by means of algebra—that is, many curves can be described and classified in terms of algebraic equations.

Descartes' work in physics turned out to be less enduring. Accepted for almost a hundred years, though mostly in France, because of his reputation as a philosopher and mathematician, his general theory of the universe, expounded in *Principia Philosophiae,* described a universe filled with vortices of matter. Originally set in motion by God, these vortices gave rise to both sun spots and the planets, which they push along in their orbits, with the earth carried around the sun by a vortex of matter with respect to which it remained stationary. This version of a heliocentric solar system allowed Descartes to dodge a personal attack by the Roman Catholic Church as being a supporter of Copernicus and to avoid the fate of Galileo. Based on nothing but philosophical speculation, these ideas lacked any kind of observational or scientifically coherent theoretical support, and they retarded acceptance on the continent of Newton's vastly superior theory. Descartes' approach, which reinforced the view of the world as a smoothly running machine, exerted a strong influence on science and philosophy in the decades to come.

Born at the Hague in the Netherlands in 1629, Christiaan Huygens, the son of a diplomat, poet, and composer, grew up in a home full of culture and tradition where Descartes was a frequent guest. After studying mathematics and law at the universities of Leiden and Breda, he decided to devote himself completely to physical science, which he did for sixteen years at home, supported by an allowance from his father. In 1666 he was invited to Paris by the Académie Royale des Sciences, where he remained and worked for the next fifteen years until his delicate health forced him to return home. After that, he only occasionally traveled abroad to meet other great scientists, including Newton. In 1695 he died at the Hague.

The work of Huygens covered a wide area, from mathematics to a variety of experimental and theoretical subjects in physics, ranging from hydrostatics to mechanics and the nature of light. While he discovered early versions of the conservation of momentum, of kinetic energy, and of the center of gravity of colliding bodies, in studying the notion of centrifugal and gravitational force he was hampered by lacking both the concept of acceleration and the powerful tool of the calculus, which had not yet been invented. His improvement of Galileo's pendulum clock has led some historians to regard him as the actual inventor of this important timepiece.

Huygens left his most permanent mark in the field of optics, in which he performed experiments and to which he contributed theoretical ideas. He and his brother Constantijn—both skillful lens grinders—constructed the best telescopes of the day, one of which he used to recognize Saturn's rings and to discover its moon Titan. He is best known for his ideas on the nature of light. In his view, light was propagated in a material medium he called ether, which consisted of an irregular conglomeration of particles that are, in a light beam, pushed forward and backward like a wave. He did not succeed in explaining optical phenomena involving polarization, however. This quite new wave theory of light was completely overshadowed by Newton's much less innovative corpuscular theory until early in the nineteenth century, when several decisive experiments determined that Huygens had been right. However, the particle-versus-wave controversy also turned out to sow the seeds for one of the ingredients of the second revolution in physical science, to which we shall turn in Chapter 10. In sum, while most of Huygens's scientific contributions served as significant stepping stones for other scientists such as Newton, his wave theory of light was a path-breaking innovation in its own right.

Empedocles had been the first to recognize that air has a corporeal substance, and some five centuries later Hero of Alexandria discovered that air could be compressed and expanded. But significant advances in our physical knowledge of air and other gases did not occur until the seventeenth century, when the Italian physicist Evangelista Torricelli (1608–1647) invented the barometer. Puzzling over Galileo's observation that no pump could lift a column of water higher than about 30 feet, he did a similar experiment with a vertical glass tube, closed on top but open at the bottom, filled with mercury. The longest column of mercury he could sustain in a glass tube closed at the top was about 30 inches. As the density of mercury is about fourteen times that of water, he concluded that, rather than nature's supposed horror of the vacuum, it must be the outside atmospheric pressure that holds up both the water and the mercury column. Moreover, he noticed that the length of the mercury column varied somewhat from day to day: the atmospheric pressure must be subject to slight variations. Thus, he had invented a way of measuring these changes—the barometer. A famous experiment confirming the notion of atmospheric pressure was performed in 1654 by Otto von Guericke (1602–1686), who used two hollow bronze hemispheres, carefully fitted edge-on-edge and evacuated by means of a pump of his own construction. Teams of eight horses, harnessed to each hemisphere and driven in opposite directions, were unable to separate them.

The next decisive step in our understanding of the behavior of gases was taken by the British natural philosopher Robert Boyle, who discovered the law that is still named after him: if temperature is held constant, the volume of a gas varies in inverse proportion to the pressure exerted upon it.[4] Boyle was a member of the Experimental Philosophy Club at Oxford, a group of intellectuals that included Christopher Wren and John Locke and which in 1662 was granted a charter by Charles II to form the Royal Society of London for Improving Natural Knowledge, known later simply as The Royal Society. Born in 1627 in Lismore Castle, the seventh son of the earl of Cork, he learned to speak Latin and French at an early age and was educated at Eton College beginning at the age of eight. After traveling in Italy as a teenager and returning to England for a fourteen-year stay at Oxford, he moved to London, where he lived for the rest of his life with his sister, supporting his scientific experiments with his own ample funds. He died in 1691.

Boyle's scientific importance rests primarily on extreme care in performing experiments, which, for him, did not merely serve the purpose of convincing skeptics but as basic tools for discovering new knowledge. In this respect he went beyond Galileo. Effectively transforming alchemy into chemistry, he was the first to recognize that chemistry was a physical science rather than just a tool for performing services and mysterious tricks. Different chemical substances, he believed, were constituted of different arrangements of corpuscles (though not of indestructible atoms), the motions of which we sense as heat. He even recognized the distinction between mixtures and compounds. This new science of chemistry would soon come into full bloom with the French scientist Antoine Laurent Lavoisier (1743–1794), who was guillotined during the French Revolution.

The most celebrated of the laboratory instruments Boyle constructed for himself, with crucial help from his assistant, Robert Hooke, was the first really effective air pump. This piece of apparatus allowed him to produce a better vacuum than anyone had before, leading him to the conclusion that Aristotle's doctrine, nature abhors a vacuum, had to be wrong. It also enabled him to perform a number of decisive experiments, one leading to the gas law mentioned above, another to the recognition that sound required air for its propagation, and yet another to the conclusion that combustion could not take place without air. A hundred years would pass before the English chemist Joseph Priestley (1733–1804) would discover that oxygen was the needed component of air.

The man who brought the first scientific revolution to its completion and at the same time personified the new physics to which that revolution gave birth was Isaac Newton. According to the Julian calendar still used in England at the time, he was born prematurely on Christmas day of the year of Galileo's death, 1642 (according to the Gregorian calendar adopted by continental Europe, his birthday was January 4, 1643), in Woolsthorpe near Colsterworth in Lincolnshire. His parents were without formal education; his father had died before he was born and his mother had remarried when he was three. Newton was brought up by his grandparents in a house without affection. The emotional damage of his childhood would be manifest in his neurotic and tortured personality for the rest of his life. At the age of twelve he was sent to grammar school in Grantham, where the curriculum consisted almost entirely of Latin and Bible studies, with very little arithmetic or mathematics. Later remembered for "his strange inventions and extraordinary inclination for mechanical works," he filled his garret room with tools and spent his time constructing waterclocks, sundials, models of machinery, and other gadgets, and became quite proficient in drawing.[5]

Thanks to the advice of an uncle and with the grudging consent of his mother, he escaped "the idiocy of rural life" and matriculated at Cambridge University at the age of eighteen, entering Trinity College as a subsizar—a student who earned his keep performing chores for the fellows and for more affluent students. The university environment surrounding him was still dominated by a stultifying Aristotelianism, but he immersed himself in the works of Galileo, Descartes, and Pierre Gassendi, the French philosopher and mathematician, a forceful advocate of a mechanistic atomism, who had died in 1655.

Newton filled his notebook with a large number of critical questions ("quaestiones") probing a variety of subjects, indicating a growing acceptance of the Cartesian mechanistic philosophy, modified by atomism (which Descartes had not accepted), and a total rejection of Aristotle. The notebook also showed that he placed great value in checking the detailed experimental consequences of any theory, and he performed some of these experiments himself, almost ruining his eyes by looking at the sun and slipping a bodkin "betwixt my eye & ye bone as near to ye backside of my eye as I could" to test his ideas on light, color, and vision. He also began to exhibit in his quaestiones a sudden great interest in mathematics, in which he was practically an autodidact, as the university almost entirely ignored that subject before establishing the Lucasian Chair in Mathematics in 1663. Only later would he study Euclid, whose style and methodology subsequently exerted a great influence on his own.

The years 1664 to 1666, often referred to as the *anni mirabiles,* were the time of Newton's most fecund scientific work, when he laid the foundations of his invention of the calculus, of the universal law of gravitation, of his laws of motion, and of his discoveries in optics. He often forgot to eat, and he slept erratically. Nevertheless, he graduated with a bachelor of arts degree in 1665. That same year Cambridge was hit by the plague, the university was closed, and he returned to Woolsthorpe for eighteen months to continue his intensely solitary work, living in a universe of his own and wanting assistance or advice from no one. Asked later how he managed to discover the law of gravitation, he replied, "by thinking on it continually."[6] What he focused on first, however, was mathematics.

In 1666 he completed three papers containing extensive original work on infinite series and the differential and integral calculus (which he called the method of fluxions), including what is now known as "the fundamental theorem of the calculus," always motivated directly by problems of motion, both rectilinear and curvilinear, which he analyzed in detail. The calculus was an independent, powerful generalization of what had been hinted at in Archimedes' method of exhaustion and in the infinitesimals of the French mathematician Pierre de Fermat (1601–1675). It would eventually blossom into a vast area of mathematics called analysis. The fundamental theorem of the calculus states that the integral and differential calculus are inverses of one another.

Shortly after returning to Cambridge Newton became a fellow of Trinity College and received the master of arts degree in 1668. The only one who knew about his path-breaking papers, which were never published, was the first incumbent of the Lucasian chair, Isaac Barrow, who promptly vacated his position so that Newton could became the Lucasian Professor of Mathematics at the age of twenty-six. He remained in Cambridge for nearly thirty years, in contact with other leading European scientists by letter and through the Royal Society, to which he was elected in 1672.

After abruptly losing interest in mathematics, Newton turned his concentrated attention to mechanics, beginning with the notion of inertia employed by both Galileo and Descartes as internal tendencies of bodies to remain in motion. Generalizing this idea, he applied it to collisions between objects and to circular motion when the centrifugal force (so named by Huygens) makes its appearance. It was also during these fertile years that he became interested in gravity—stimulated, according to his own account, by watching an apple fall from a tree. Could the same force that made the apple fall extend all the way into space and make the moon "fall" around the earth? If the moon would fall, pulled by the same force as the apple, he conjectured, it was the centrifugal force that would balance this pull and keep the moon in its orbit. That insight did not just come in a flash of intuition; it took him another twenty years to work it all out and to complete these studies, which he had begun between 1664 and 1666.

Meanwhile, Newton was also intensely engaged in optical experiments and in unorthodox theological speculations in sympathy with Arianism, risky though such thoughts were for his future at Cambridge, as well as in alchemy. He had been introduced to chemistry by Robert Boyle, with whom he corresponded until the latter's death, but he explicitly gave it up and turned to alchemy, which he regarded as more profound.

Experimenting with prisms, Newton discovered that the white light of the sun actually consisted of a mixture of colored components. He had found that the index of refraction of glass varies with the color of the light shining through it, so that in a prism, different colors were bent by different angles (Fig. 9). Taking advantage of this fact, he was able to separate the various constituents of white sunlight by allowing it to pass through a prism, so that the colors in it emerged at distinct angles. Directing all these separate pencils of colored light through another prism, he was also able to reconstitute them back into a single ray of white light. Other experiments with thin films and circular lenses placed on flat panes of glass led him to an explanation of the resulting multicolored circles; the colored rings we see when sunlight falls on oil films floating on water are still called Newton's rings.

Another outcome of his optical experiments was the invention and construction of the first reflecting telescope, in which the light, instead of passing straight through several lenses before emerging at the other end, is internally reflected by a concave mirror. This design allows the telescope to be very much shorter and less cumbersome as an instrument for observing the sky. Newton's ideas on the corpuscular nature of light, on the other hand, would mislead the scientific world for almost two hundred years. Contrary to Huygens's view, which would be proved essentially correct early in the nineteenth century, he saw light as made up of a stream of small particles.

Newton's greatest work was the monumental *Philosophae Naturalis Principia Mathematica,* which he presented to the Royal Society in 1686 but which, because of a shortage of funds, could not be published until 1687, after his friend, the astronomer Edmund Halley (1656–1742), strongly pushed for—and paid for—its printing. The book contained his two most important contributions to physics, the

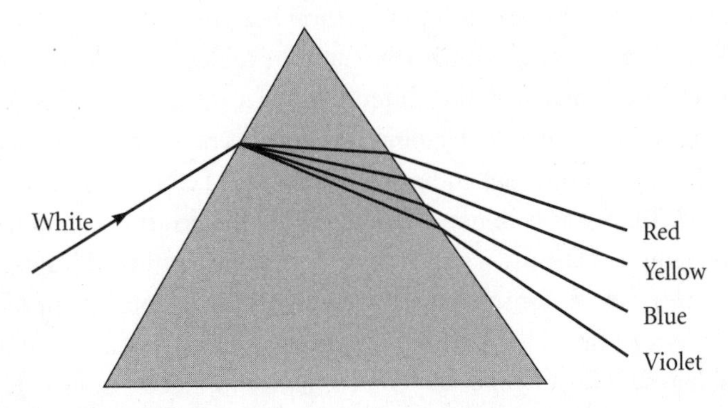

Figure 9 The separation of the colors in white light by a transparent prism, discovered by Isaac Newton.

laws of motion and the universal law of gravitation. After its publication, which caused a sensation in Europe, Newton lost interest in these subjects and devoted himself to more mundane matters, such as being elected a member of Parliament representing Cambridge University. He also became involved in a number of unpleasant, protracted priority disputes that revealed the prickly side of his character, particularly his extreme sensitivity to any hint of criticism.

The best known of these quarrels involved the German philosopher and mathematician Gottfried Wilhelm Leibniz, who was born in Leipzig in 1646 and entered its university at the age of fifteen. There, he studied law, philosophy, and Euclidean geometry, and, on his own, the scientific works of Galileo, Kepler, Descartes, and others. Receiving his doctorate in law at the age of twenty, he traveled about Europe and soon entered the service of the elector and archbishop of Mainz, who employed him for diplomatic missions and the devising of plans to preserve the peace in Europe. At the invitation of King Louis XIV he went to Paris, where he stayed about three years and became seriously interested in mathematics. In 1672 he constructed a calculating machine that could multiply, divide, and extract roots. After the death of the elector of Mainz, a visit to London deepened Leibniz's acquaintance with the work of Isaac Newton as a mathematician and led to his election to the Royal Society.

As he began to work on mathematical problems, in 1675 Leibniz independently invented the differential and integral calculus, employing a notation and terminology that were much more user-friendly, as we would say today, than Newton's; Leibniz's system won out and still remains in use. After accepting the position of librarian to the House of Brunswick in Hanover, while he continued his diplomatic missions, Leibniz was instrumental in the establishment of an academy of science in Berlin in 1700, of which he was appointed president for life. As imperial privy counselor in Vienna from 1712 to 1714, he spent the last years of his life embittered by his quarrel with

Newton and suffering from gout. He died in Hanover in 1716, ignored by both the Berlin Academy and the Royal Society.

The dispute over the invention of the calculus arose primarily because Newton's original paper of about 1665, seen by Barrow only, had remained unpublished, and Newton continued to be neurotically uncommunicative and even secretive about his work. When Leibniz invented the same mathematical method ten years later and published it, at a time when Newton's interests had shifted to other things, Newton (whose paranoia was as active as ever) publicly accused Leibniz of plagiarism. Leibniz's appeal to the Royal Society was adjudicated by a committee appointed by Newton, who then wrote its "impartial" report himself and even anonymously reviewed it. The result of this bad treatment of Leibniz, justifiably viewed on the continent as blatantly unfair, was a long-lasting isolation of British mathematics from the rest of Europe, to the detriment of both.

Another great feud involving Newton was his drawn-out controversy with Robert Hooke (1635–1703), one of the founders of the Royal Society and its long-time curator and secretary. Hooke was a versatile scientist of great distinction, whose temperament was as disputatious as that of Newton. His *Micrographia*, published in 1665, dealt with the nature of light (he agreed with Huygens's view that light was a wave phenomenon), optics, and microscopy, including many improvements of the microscope. It contained, as well, a number of remarkably detailed drawings of microscopic animals, made from his own observations. His most significant contribution to physics was the discovery of the proportionality of strain and stress in an elastic body (still called Hooke's law), which was the reason why a spring vibrates with a regular period, just like a pendulum.

The source of Hooke's arguments with Newton—just one of many priority disputes Hooke vigorously pursued—was his work on gravitation, an idea that had occurred to many other scientists at the time. Hooke had even played with the notion that the force of gravity obeyed an inverse-square law that was in some way responsible for the orbits of the planets. When Newton, fully aware of this work, published his universal law of gravitation without any credit or reference to it, Hooke accused him of plagiarism. (In his biography of Newton, *Never at Rest,* Richard Westfall argues in detail that, although the ideas Hooke expressed in his correspondence with Newton sounded superficially as though they anticipated Newton, they were quite confused, in contrast to those of Newton, which were very precise.)

The intensity of his concentration and the stress produced by his various recurring disputes finally began to have an effect on Newton's health, and in 1692 he fell into a severe, disabling depression that lasted for several years. After his recovery, he no longer pursued his work in physics and mathematics with his previous zeal, and instead accepted an appointment as warden of the Mint, becoming master in 1699. He took this position very seriously, devising new methods to prevent forgery and clipping of coins. In 1703 he was elected president of the Royal Society, the following year he published his *Opticks,* which contained all the results of his optical research, and the year after that Queen Anne knighted him.

Sir Isaac spent the remaining years of his life revising the *Principia,* and when he died in 1727 he was accorded a state funeral at Westminster Abbey "like a king," as Voltaire remarked. To this day, judgments of Newton as a scientist differ drastically from judgments of him as a human being. Aldous Huxley put it succinctly: "The price Newton had to pay for being a supreme intellect was that he was incapable of friendship, love, fatherhood, and many other desirable things. As a man he was a failure; as a monster he was superb."[7]

The two great theories contained in the *Principia* are Newton's laws of motion and his law of universal gravitation. The first is intended to govern the way all physical objects move under the influence of given forces of any kind; the second describes the action of one particular force, that of gravity. The laws of motion comprise the following three specific rules. The first states that any object, left undisturbed and not subject to any force, will remain either at rest or in a state of uniform rectilinear motion forever. The second, based on the stipulation that every physical object has a property called mass, says that an object of mass m, subject to force F, will experience an acceleration a in the same direction as F and of magnitude $a = F/m$. (The law was actually explicitly stated in this form for the first time by Euler in 1750.)[8] The third law dictates that if object A exerts a force on object B, then B exerts an equal force on A in the opposite direction.

Newton's laws of motion build on Galileo's recognition of the crucial importance of the role of acceleration as well as Huygens's idea of momentum conservation and even on Buridan's notion of impetus. As Newton himself had declared, "If I have seen farther, it was by standing on the shoulders of giants." (This famous aphorism seems to have been a standard phrase of long standing, and Newton's intended meaning is controversial—he may have meant it as a sarcastic dig at the plagiarism accusations of Robert Hooke, who was of small stature and somewhat deformed.)[9] The law of universal gravitation says that gravity is a force of attraction that acts between any two objects of masses m and M a distance D apart, attracting the mass m to M and the mass M with equal force to m; the magnitude of that force is proportional to mM/D^2.

The power of these detailed laws consists in their universal applicability. Assuming that we know all the forces, if any, in addition to gravity (for example, Hooke's spring forces) acting between any given group of physical objects, they, together with Newton's calculus, allow us in principle to calculate the future behavior of all of them if we know their positions and velocities now. Newton himself applied these laws to the sun, the earth (including the tides), the other planets, and the moon, as well as to comets, to determine their orbits in the solar system, assuming that the only force acting among them is that of gravity, and he managed to derive from them all of

Kepler's laws. The division of the laws underlying the movements of a group of bodies into general laws of motion on one hand, applicable no matter what the forces on or among them may be, and laws that govern the action of specific kinds of forces on the other hand has had an enormous impact on all of physics to this day, with the exception of Einstein's general theory of relativity. Forces that Newton knew nothing about have been discovered since, but his laws of motion are still regarded as valid, except when velocities approach the speed of light or when the objects are of submicroscopic size. When NASA sends a probe to Mars, it relies on Newton's laws of motion, as well as on his law of gravity, in calculating the space capsule's trajectory.

Whereas the general response among physicists to the laws of motion was quite positive, the reaction to Newton's law of gravitation was revulsion, especially from Leibniz. Lacking an explanation of how one object managed to exert a force on another without touching it, his "action at a distance"—postulated and described purely mathematically in terms of its dependence on how far apart two objects are from one another—was universally regarded as repugnant, even initially by Newton himself. Its coldly abstract form differed dramatically from the more intuitive physical picture of the vortices Descartes had postulated. Though gravitation was gradually accepted, the time would come, some 150 years later, when the new idea of a "field" enabled physicists to eliminate action at a distance, rendering that concept once again taboo.

As a consequence of Newton's achievement, physicists would be able to look at the entire universe, in microcosm as well as macrocosm, as a giant machine not just in a metaphorical sense but in the more literal sense that they could imagine themselves actually being able to predict in detail the future course of events, both in the heavens and here on earth. Newton, of course, did not have full knowledge of all the physical processes that would enable anyone to imagine such predictions could be realized. It took until the end of the nineteenth century to acquire even a semblance of the needed knowledge. But it was through his unification of celestial and terrestrial mechanics, together with both his thorough-going mathematization of physical science and his invention of the required mathematical tools, that a dream of such predictive power would not seem totally far-fetched.

We should not conclude from this that Newton himself or his contemporaries interpreted his laws as implying a machinelike universe running on its own forever once set in motion by God. In a famous exchange of letters between Leibniz and Samuel Clarke, a royal chaplain and a supporter of Newton who was in touch with the master, Leibniz accused Newton of expecting occasional intervention by God in the running of the universe, like a clockmaker who needs to adjust his mechanism from time to time.[10] It was particularly Newton's expressed view of space as the "sensorium of God" that Leibniz could not swallow: he saw that as opening the world up for miracles. By contrast, Leibniz's view was that the cosmos ran like a well-oiled machine, perfectly constructed by God and never in need of any further tinkering. Such arguments made Leibniz vulnerable to accusations of materialism, as his clockwork universe had no need of God. Thus, though it was Newton who made it possible to regard the world as a smoothly running machine, Newton himself was still too much of a "magician"—in John Maynard Keynes's phrase—to see it that way himself.

five

Newton's Legacy

With the death of Isaac Newton, the first scientific revolution ended and physics entered a new era, distinguished by two dominant characteristics: a strong emphasis on experimental evidence for theories and a penchant for formulating these theories in mathematical language. During the next two centuries, progress in physical science also included new observations in astronomy that changed much of our view of the universe. This progress fell into two categories. The first comprised a fleshing out and enormous further evolution of the calculus and its consequences for the laws of mechanics. The second consisted of the discovery of new physical phenomena and the confirmation they provided for such old conjectures as atomism, as well as the invention of novel theories to explain them. The areas that primarily benefited from this progress were those of sound, light, electricity, magnetism, and heat, the last contributing most directly to the next scientific upheaval.

In astronomy, three observers added a large number of new and very precise observational facts to our knowledge of the universe, some of which would play important roles both in verifying the accuracy of the Newtonian laws and eventually in demonstrating their limitations. Frederick William Herschel was born in 1738 in

Hanover, Germany, and spent his early life as a musician, playing the oboe in a military band. He emigrated to England at the age of nineteen, earning a living as a conductor, composer, organist, and music teacher. Gradually developing an ever more intense interest in astronomy, he began grinding lenses and mirrors and building telescopes, greatly assisted by his sister, whom he had fetched from Hanover. His serious astronomical work began in 1773, and he created a sensation in 1781 by discovering the first new planet since antiquity. Though he wanted to call it Georgium Sedum in honor of King George III, the name Uranus was ultimately chosen for this seventh planet in the solar system. Later, he also discovered two satellites of Uranus (Titania and Oberon), as well as two of Saturn (Enceladus and Mimas).

One of Herschel's lasting contributions consisted of two systematic surveys of the whole sky, which he carried out by means of a large reflector telescope of his own construction, the best anywhere to be found at the time. His attention was particularly focused on double stars—two stars that appear to be very close to one another—of which he assembled three catalogues, listing a total of 848 such pairs. He discovered that the relative motion of some of them could be explained only by assuming they were rotating about one another, evidently held together by gravity—the first appearance of that force outside the solar system. The resolving power of his telescope enabled him to determine that the indistinct stellar objects called nebulae were in fact large conglomerations of stars—galaxies—which he assembled in two systematic catalogues, the first listing 2,500 and the second 5,000. The existence of such large numbers of galaxies led him to formulate the first scientific theory of the development of the universe: it began as a disorganized distribution of stars, which eventually clumped together, forming galaxies. Herschel may thus be regarded as the father of modern observational cosmogony—the study of the origin and history of the universe.

From observations of the sun, Herschel was able to draw two important conclusions. The first originated from his use of a variety of darkening glasses of different colors for solar observations. Since some of these gave him a distinct sensation of warmth while others did not, he concluded that the heat must be caused by some of the radiation emanating invisibly from the sun. He thus discovered the heat radiation that we now call infrared. Second, he demonstrated that the data published by others of varying solar positions relative to the stars made sense only if interpreted as originating from a proper motion of the sun itself. Thus, the sun was displaced from the center of the universe and, by implication, the position of the earth was further diminished.

Herschel continued to build telescopes for sale until he married a wealthy widow, with whom he had a son, John, who subsequently became a prominent astronomer in his own right, carrying on the legacy of his father. Herschel was awarded many honors, appointed court astronomer, and knighted in 1816. He died peacefully in 1822 in Slough.

The second astronomer of note, though of lesser stature than Herschel, was Heinrich Wilhelm Matthäus Olbers (1758–1840), who was born and died in Bremen, Germany. His main interest was in asteroids and comets, but he is remembered primarily for formulating a paradox that continued to puzzle astronomers long after his death and would not be resolved for 150 years. Olbers's paradox is this: if the universe is infinitely large and more or less uniformly populated with stars, all shining like the sun, why is the night sky dark, rather than brightly lit at every point by this infinite glow? The only way he was able to explain this was by assuming that interstellar space was not entirely transparent but absorbed a small amount of starlight, so that we can see the nearby stars against a dark background. This resolution of Olbers's paradox was incorrect, but an accurate solution would have to await the discovery of red shift and the expansion of the universe in the twentieth century.

Friedrich Wilhelm Bessel, the third of our trio of astronomers, was born in 1784 in Minden, Germany. His paper on Halley's comet, written at the age of nineteen, so impressed Olbers that he arranged a position for the young man at Lilienthal Observatory. Four years later he was commissioned by the Prussian government to construct a large observatory in Königsberg, where he was subsequently appointed professor of astronomy and remained until his death in 1846. Tabulating the positions of about 50,000 stars, Bessel set new standards for their position measurement. He assembled personal statistics on the idiosyncrasies of each observer and made a systematic study of instrumental errors, taking into account not only the technical imperfections of his own telescope but also external disturbances in the atmosphere.

Knowing the angular positions of distant stars precisely allowed Bessel to calculate the distances of the nearer ones by very accurately measuring the parallax caused by the motion of the earth in its orbit around the sun; their positions relative to the background of faraway stars would change in the course of half a year, facilitating the exact measurement of their parallax. In another major discovery, he found two bright stars, Sirius and Procyon, that had minute motions of their own, which indicated to him that they must each have invisible partners with which they interacted gravitationally, executing a ghostly dance like Herschel's double stars. After his death, these companions of Sirius and Procyon were indeed found by means of more powerful telescopes. For the convenience of his detailed calculations, Bessel defined a large set of functions, now known as Bessel functions, that are still indispensable as tools in mathematical physics and chemistry.

While alive and active, Isaac Newton had been able to defeat all his questioners and adversaries by his sheer intellectual power as a superior mathematician wielding a new weapon of his own invention, the calculus. This combination of talent and tool enabled him to calculate the precise consequences of his law of gravity, one of which was to prove Kepler's laws. All the revulsion for action at a distance notwithstanding, there was no denying that the law of gravity worked, as he could convincingly demonstrate. However, the calculus proved to be a mathematical method of vastly greater potential than either Newton or Leibniz initially envisioned, and it took most of the eighteenth and nineteenth centuries for mathematicians to exploit it fully and plumb its depths.

In the meantime, the laws of mechanics also needed to be applied to the explanation of the behavior of fluids, which after all, if the atomists were right, were made up of particles that ought to obey Newton's laws. The first modern work on fluids, or hydrodynamics, was done by the Swiss mathematician Daniel Bernoulli, the second son of the mathematician Jean Bernoulli (1667–1748) and nephew of Jacques Bernoulli (1654–1705), also a mathematician. Both Daniel's father and his uncle made important contributions to the calculus and to the theory of probability. Born in 1700 in Groningen, the Netherlands, Daniel was appointed to the chair of mathematics at the St. Petersburg Academy of Science in 1725, but eight years later he returned to Switzerland and moved to the University of Basel, where he remained until his retirement; he died in Basel in 1782.

Bernoulli was a polymath, making significant contributions to astronomy, mathematics, and physics, among other fields. His work on gases and liquids was both practical and theoretical, based on the assumption that these fluids were composed of tiny particles. He found an equation that determined the dependence of atmospheric pressure on altitude; and as a consequence of the conservation of energy, he derived the principle that still bears his name: the pressure in a fluid decreases as its velocity increases. This principle is responsible for the lift a properly shaped airplane wing produces as it forces the air on its upper side to move faster than on its underside. He also made important advances in the theory of partial differential equations—an outgrowth of the calculus that turned Newton's mechanics into an engine for large-scale predictions of the motions of interacting bodies. Together with Leonard Euler, whom he had met in St. Petersburg, he also did extensive work on the differential equations governing vibrations.

Leonard Euler was born in Basel in 1707, son of a Protestant minister with a lively interest in mathematics who had himself attended lectures by Jacques Bernoulli. With a prodigious capacity for mental calculations and great scientific curiosity, Euler studied under Jean Bernoulli in Basel and earned his master's degree there in 1723 but could not find a position in Switzerland. Invited by Daniel Bernoulli to St. Petersburg, he obtained a post as professor of physics and, when Bernoulli left St. Petersburg, Euler succeeded him there as professor of mathematics. In 1744, having come to Berlin at the request of King Frederick the Great of Prussia, he was appointed director of the Berlin Academy of Sciences, but returned to St. Petersburg in 1766 at the invitation of Catherine the Great to become director of the academy there. Shortly after that, a cataract in one of his eyes left him totally blind, since he had earlier lost the sight in his other eye by looking at the sun for extensive periods of time, pursuing astronomy. He died in 1783.

Euler contributed significantly to all the classical fields of mathematics, especially those relevant to physics. Formulating the calculus in algebraic terms rather than in geometric ones as Newton had done, he cast it in the form we employ today. His use of differential equations in celestial mechanics and in particular for the analysis of the motions of the earth and the moon led him to a detailed explanation of the rhythm of the tides, and from there to the behavior of fluids. Many of his results remain as permanent pieces in the foundation of mathematics and mechanics.

Seven other great mathematicians who overlapped from the eighteenth to the nineteenth century followed a similar path of working in, and notably advancing, both the calculus and its application in Newtonian mechanics or other areas of physics. They were Karl Friedrich Gauss of Germany and six Frenchmen: Jean le Rond d'Alembert, Joseph Louis Lagrange, Pierre Simon de Laplace, Adrien-

Marie Legendre, Jean Baptiste Joseph Fourier, and Augustin-Louis Cauchy. During the nineteenth century their successors along this line were the Scot William Rowan Hamilton (1805–1865), the German Karl Gustav Jacobi (1804–1851), and the Frenchman Joseph Liouville (1809–1882), as well as Jules Henri Poincaré, whose life extended into the twentieth century. In addition, the work of the nineteenth-century pure mathematicians Évariste Galois and Bernhard Riemann turned out to have seminal consequences for physics.

Jean le Rond d'Alembert was found on the doorstep of a church in Paris shortly after his birth in 1717, but the Chevalier Destouches-Canon, assumed to have been his father, took care of his financial needs during childhood; his mother was a courtesan. He was schooled in law and medicine at the Jansenist Mazarin College but soon abandoned these fields and turned to mathematics, which he pursued for the rest of his life, making important contributions to the development of the calculus and its applications to dynamics.

The problem in celestial mechanics that fascinated d'Alembert in particular was that of three bodies subject to mutual gravitational attraction. Newton had completely solved the two-body problem: the earth or another planet and the sun, or the earth and the moon, attracting one another by gravity. But when three bodies are involved simultaneously, the problem of determining their motion becomes so difficult that it would challenge and frustrate many great mathematicians long after Newton; and indeed, it is not completely solved to this day. However, as often happens in mathematics, even when a difficult problem defies solution, the attacks on it generate new methods that turn out to be very useful in other contexts. D'Alembert's ideas fertilized the entire area of differential equations, and he used some of them to put Newton's notion of the precession of the equinoxes on a sound mathematical foundation, explaining the motion of the earth's axis with respect to the ecliptic. D'Alembert became a member of the Encyclopedistes, contributing articles on scientific topics and acting as an editor. He died in Paris in 1783.

Joseph Louis Lagrange, born in Turin, Italy, in 1736, of a French father and Italian mother, was almost entirely self-taught in mathematics but showed such formidable powers that he was appointed professor of mathematics at the Royal Artillery School in Turin at the age of nineteen. There he remained until 1766, when he was invited by Frederick the Great to succeed Leonard Euler as director of the Berlin Academy of Sciences. In 1787 King Louis XVI invited him to come to Paris as a member of the French Royal Academy, and ten years later he was appointed professor of mathematics at the École Polytechnique.

Lagrange's early work on an important calculus problem that had long stymied Euler so impressed the elder man that he delayed the publication of his own solution until Lagrange's discovery was in print. Lagrange then began to think about the celestial three-body problem that d'Alembert had tried to tackle. Though not completely successful in solving it, he won the Grand Prix of the French Academy for his contribution to its solution, and he won the same prize again two years later for his work on the even more complicated problem of Jupiter and the four satellites discovered by Galileo. He then assembled all his contributions to the use of differential equations in Newtonian mechanics in his great treatise *Mécanique Analytique.* The phrase *analytical mechanics* is still used for that subject, and his approach to the mathematical formulation of equations of motion remains very influential to this day. He was particularly proud of the fact that his entire tome did not contain a single diagram; in contrast to Newton, he worked it all out algebraically.

Later in his life, Lagrange was appointed president of a commission to standardize weights and measures, where he successfully advocated the universal use of base 10 rather than 12, thereby becoming the father of the metric system now used in most of the world outside the United States. He died in Paris in 1813 and was given the special honor of being buried in the Panthéon.

The mathematician Pierre Simon de Laplace represents the personification of deterministic physics, as he is the author of the famous declaration, "Given for an instant an intelligence which could comprehend all the forces by which nature is animated . . . an intelligence sufficiently vast to submit these data to analysis—it would embrace in the same formula the movements of the greatest bodies and those of the lightest atoms; for it nothing would be uncertain and the future, as the past, would be present to its eyes."[1] Here we have the universe as clockwork in a nutshell. However, this very statement was made in a context justifying the use of probabilistic procedures, to which Laplace made seminal contributions. A perfect clockwork though the world was, we lack the detailed knowledge needed to exploit its machinelike constitution for telling its future; therefore, we have to resort to statistics and probabilities for our less-than-perfect predictions.

Born in 1749 in Beaumont-en-Auge, Normandy, Laplace was educated from age seven until he was sixteen at the Benedictine College in Beaumont-en-Auge, destined for a career in the Church. Instead of following that path, he entered Caen University and after two years went on to Paris with a letter of recommendation to d'Alembert, whom he quickly impressed as a promising mathematician and who arranged for him to be appointed as a professor at the École Militaire. At the age of twenty-four he was elected to the Royal Academy of Sciences, where he eventually became one of the leading members. Widely disliked by other scientists, ambitious and offensive at times to many, including both Legendre and Lagrange, Laplace was nevertheless recognized as a mathematician peerless in his ability to extend and generalize ideas, often those initiated by others. After narrowly escaping the fate of Lavoisier during the Revolution, he was appointed by Napoleon first as his minister of the interior and then as a senator. However, he later voted to overthrow Napoleon in favor of the Bourbon monarchy, to which he remained loyal for the rest of his life. After the restoration of the monarchy,

Laplace was rewarded by being made a marquis. He died in Paris in 1827.

The first set of problems that attracted Laplace's interest were all connected with celestial mechanics. To the great puzzlement of astronomers, it had been observed that the orbit of Jupiter seemed to be continually shrinking while the orbit of Saturn appeared to be expanding. Was there something wrong with Newton's law of gravitation? To the contrary, Laplace explained the phenomenon perfectly within the framework of Newton's theory. Owing to the interaction of the two planets and the fact that their periods were almost commensurable, the size of their orbits could be expected to change periodically, with a period of 929 years. He next attacked the problem of the exact motion of the moon around the earth, which had frustrated both Euler and Lagrange. The moon's orbit is determined principally by the moon-earth attraction, but the pull of the distant sun also plays a minor role, which changes with time as the orbit of the earth responds to the perturbing presence of the other planets. As a result, the mean motion of the moon slowly speeds up while that of the earth slows down, with a period of some millions of years.

This was Laplace's first application of a mathematical method of his invention, called perturbation theory, which, despite several numerical errors in his calculations, would turn out to be enormously useful in many other contexts in physics. When a problem that involves relatively weak perturbing influences on a dominant system is too difficult to solve exactly (such as the earth and its close-by moon being disturbed by the far-away sun and other planets), perturbation theory can often be used to find an excellent approximation to a solution. The method served him well for finding the answer to another celestial problem that worried him: can we be sure that the planetary system as a whole, with all the gravitational pulls of the planets on one another, is really stable? Is there a chance that the entire assembly will eventually break up, with each planet flying off in a different direction, some perhaps crashing into the sun? To his relief, he could prove that the system was stable and there was no need to worry. While this conclusion was true to the extent that perturbation theory was a reliable approximation, and the solar system will indeed be stable for a very long time, we now know that eventually it will break up, owing to the mutual interactions of all its constituents.

Laplace was also able to calculate with great precision both the ebb and flow of the tides and the precession of the equinoxes. The elaborate theory he devised for the formation of the solar system, however, turned out to be flawed; it could not correctly account for some of the properties of the sun and the planets. In addition to his great work in astronomy, Laplace made deeply influential contributions to all parts of analysis, and to the development of partial differential equations in particular. To this day, there is hardly an area of mathematical physics that does not make use of a technique invented or a result obtained by Laplace.

Adrien-Marie Legendre, born in Paris in 1752 and educated there at the Collège Mazarin, did not impress the world of mathematics until the age of thirty, when he won a prize awarded by the Berlin Academy of Science. A year later elected to the French Academy, he began to publish important papers in mathematics and became a member of the Royal Society. After the Revolution, which interrupted his work for a while, he was appointed professor of mathematics at the Institut de Marat in Paris and head of the department set up to standardize weights and measures on a decimal basis as decreed by the revolutionary regime. In 1813 he succeeded Laplace as the head of the Bureau de Longitudes, a position he retained until his death in Paris in 1833.

Though Legendre's stature as a mathematician is somewhat lower than that of the others in this group, he made important contributions to number theory, elliptical functions, the calculus of variations, and celestial mechanics. In the physical sciences and astronomy he is mostly remembered for his introduction of the Legendre polynomials, solutions of a differential equation named after him. These polynomials have turned out to be extremely useful in many areas of physics.

Jean Baptiste Joseph Fourier is the fifth of the eighteenth-century French mathematicians relevant to this story of how the calculus developed into an indispensable tool for all physical science. He was born in Auxerre in 1768, the son of a tailor but orphaned when quite young; he was educated at the local military academy and subsequently at a Benedictine school. After the outbreak of the Revolution, he was arrested for defending victims of the Terror but was released after Robespierre was executed. By that time he had discovered his strong interest in mathematics, and he went to Paris to study at the École Normale. Soon he became a lecturer at the École Polytechnique. In 1798 Napoleon selected him to come along on his Egyptian campaign, where he performed diplomatic missions. Upon his return to Paris, Napoleon appointed Fourier prefect of Isère in southern France and later prefect of the Département du Rhône, administrative positions for which he showed a strong aptitude. As a reward for his services, Napoleon made him a baron and subsequently a count.

All the while, Fourier did mathematics in his spare time. However, in protest against the emperor's autocratic rule during the hundred days after his return from Elba, he resigned his post and managed to be appointed head of the Bureau of Statistics, where he could pursue mathematics full time. He was elected a member of the French Academy of Sciences, of which he became the permanent secretary, and the Académie Française, as well as a foreign member of the Royal Society. He died in 1830 as a result of a disease contracted while in Egypt.

Fourier's primary mathematical interest was in the area of partial differential equations, to which he made a contribution that has become a crucial tool in all fields of physical science, especially in physics. He discovered that every periodic function (a function that repeats itself over and over) can be analyzed and expressed as an infinite sum of trigonometric functions, that is, sines and cosines. If the solution of a linear partial differential equation—these were ubiquitous in all the newly emerging areas of physics—is written in the form of such a Fourier series, each term in the sum separately satisfies an equation that is much easier to solve than the original one. The fields of electromagnetism and quantum mechanics are unthinkable without Fourier analysis.

This brings us to the man often called the prince of mathematicians. Karl Friedrich Gauss was born in 1777 in Braunschweig, Germany, to a poor and uneducated family. After teaching himself to read and count, he was spotted at the age of eight by his grammar school teacher as an outstanding mathematical talent. The teacher had assigned his class the task of adding up all the numbers from 1 to 100, expecting to have peace and quiet for some time. However, young Karl presented him with the desired sum S within a few minutes: he had written all the numbers once in their regular order and underneath in reverse order. Since each of the resulting 100 columns of two numbers added up to 101, he concluded that if he multiplied 100×101, the result would have to be twice the sum S, once for each of the two rows. In other words, at the age of eight he had found the formula $1 + \ldots + 100 = (100 \times 101)/2$. The teacher concluded that there was little he could teach his young pupil, and Karl should be educated for a profession.

At the age of fourteen he was presented to the court of the duke of Brunswick, whom he so impressed with his calculating talent that the duke supported him with a generous allowance until his own death in 1806. He attended the University of Göttingen from 1795 to 1798, and by the time he received his doctorate at the University of Helmstedt, he had already made most of the many fundamental discoveries in mathematics for which he is famous. In his doctoral dissertation he proved what is known as the fundamental theorem of algebra, which states that every algebraic equation with complex coefficients has at least one root that is a complex number. (A complex number is a number of the form $a + ib$, where i is the "imaginary number" $\sqrt{-1}$, and a and b are ordinary real numbers. Imaginary numbers were not yet fully recognized by all mathematicians at that time, but today they are indispensable tools in all physical sciences and engineering.) Gauss also invented the now universally employed system of representing complex numbers as points in a plane. He then developed an interest in astronomy and was appointed not only professor of mathematics but also director of the Göttingen observatory, a position he retained for the rest of his life.

Although Gauss was recognized early by European mathematicians, made a foreign member of the Royal Society in London, invited to join the French and Russian academies of sciences, and offered a position in St. Petersburg, he never had a happy private life. After suffering as a youth from being brought up by a difficult father, his relationships with his own sons were less than cordial. And owing to the political turmoil of the Napoleonic wars, his financial situation was always precarious. Politically very conservative and unsympathetic to the contemporary revolutions in Europe, he was regarded as cold and distant by those who knew him. Much of his groundbreaking work remained unpublished until after his death, when many other mathematicians discovered that, unbeknown to them, he had anticipated their results. There is hardly any part of mathematics or its applications to physics to which Gauss did not make basic, indispensable contributions.

While Gauss's mathematical work habits were solitary, he developed a long-lasting friendship with Alexander von Humboldt that left its imprint on German science, and he had many active correspondents in astronomy, a field he avidly pursued by scrutinizing the heavens. His method of calculating the orbit of an asteroid or planet on the basis of only three observations remains a classic. (Long after his death, the 1001st planetoid to be discovered was named Gaussia.) He also did pioneering research in crystallography, optics, and mechanics as well as electricity and magnetism. In recognition of his work on magnetic phenomena, a unit of magnetic flux density is named after him. At the age of 62 he added to his linguistic accomplishments by learning Russian, and in 1855 he succumbed to heart disease in Göttingen at the age of 77.

Twelve years younger than Gauss, Augustin-Louis Cauchy was born in 1789 in Paris. To escape the Terror, his family moved to the village of Arceuil, where Pierre Laplace turned out to be their neighbor. Receiving his early education from his father, a prominent lawyer and classical scholar, he also came to the attention of Lagrange, who recognized his talent but, as the story goes, urged his father to give the boy a good classical, literary education and not to let him look at a mathematics book until the age of seventeen. This introduction to mathematics finally took place at the École Polytechnique in Paris, after which he studied engineering at the École de Ponts et Chausses. For the next four years, he worked as a civil engineer and subsequently returned to Paris, where, at the age of 27, he became a professor at the École Polytechnique. That same year, the Bourbon monarchy was restored, and some prominent scientists who had been republicans and Bonapartists were expelled from the Academy of Sciences. As a replacement, Cauchy was appointed to the academy, having just won its Grand Prix for one of his mathematical papers. This marked the beginning of a very productive mathematical career, and he was soon appointed to a chair at the Collège de France.

Like many other French scientists, Cauchy suffered the vicissitudes of the political turmoil brought about by the Revolution and its aftermath. Because he refused to take the new oath of allegiance after the overthrow of the Bourbon king Charles X in 1830, he was forced to resign his professorship. He had meanwhile married and become the father of two daughters, whom he had to leave when he went into exile in Fribourg, Switzerland, to live among the Jesuits. His next stop was the University of Turin, where he was appointed to the chair of mathematical physics. Joined by his family, he served as tutor of the son of Charles X in Prague from 1833 to 1838, with a barony as a reward. In 1838 Cauchy finally returned to Paris and to his professorship at the École Polytechnique. From 1848 to 1852 he served as a professor at the Sorbonne and died at Sceaux, near Paris, in 1857.

Cauchy's best work was all published in the 1820s, in the form of three treatises: *Cours d'Analyse de l'Ecole Polytechnique; Résumé des leçons sur le calcul infinitésimal;* and *Leçons sur les applications de calcul infinitésimal à la géométrie.* His principal contribution consisted in making the basis of the calculus mathematically clean and rigorous, and he was the first to give a proof of the Taylor expansion of a function. The Taylor expansion of a function of x expresses that function in the form of an infinite sum of successive powers of x, a form that has found a great variety of applications in most areas of science. In the *Cours d'Analyse* he provided the first general analysis of functions of complex numbers. Cauchy is regarded by many as one of history's greatest mathematicians.

By the middle of the nineteenth century, mathematical techniques for the calculation of planetary orbits using Newton's law of gravity and his laws of motion—at this time these were usually expressed in the form given to them by Euler, Lagrange, or the French mathematician and physicist Siméon-Denis Poisson (1781–1840), or else in the alternative forms invented by Hamilton and Jacobi—were so robust and reliable that any slight discrepancy between predictions and observations were regarded as serious scientific problems. One such discrepancy was the fact that the major axis of the elliptical orbit of the planet Mercury appeared to be rotating at a rate of 38 seconds per century. No satisfactory explanation of this phenomenon could be found, and it was left as an unsolved puzzle until early in the twentieth century, when its solution would be provided by Einstein's general theory of relativity.

The second discrepancy concerned the planet Uranus, which Herschel had discovered in 1781 and for whose motion the mathematical astronomer Alexis Bouvard (1767–1843) had calculated precise tables in 1821. However, by the mid 1840s these tables were already found to be grossly inaccurate. When the French astronomer Urbain Jean Joseph Leverrier (1811–1877) learned about this failure of astronomical predictions, he suggested that the cause of the disturbance of the orbit of Uranus may be the attraction of another planet, yet unknown, the existence of which had already been conjectured by Herschel. He even went so far as to use the observed perturbation of Uranus to publish a calculated prediction of both the position and size of the unknown object. Sure enough, the astronomers Johann Galle and Louis d'Arrest at the Berlin Observatory promptly found the new planet within 1° of Leverrier's prediction; since it appeared to have a greenish hue, it was named Neptune. Because a young English astronomer, John Couch Adams, had already made the same calculation as Leverrier and sent it to George Airy, the astronomer royal, who failed to read it until he saw Leverrier's paper in print, a priority dispute arose, which was aggravated by chauvinism in both Britain and France. The name Neptune rather than Leverrier was chosen as a compromise.

The last of the great mathematicians to contribute decisively to the development of the deterministic system of Newtonian mechanics was Jules Henri Poincaré. At the same time, he also discovered that this system was not quite as deterministic as had been thought. Poincaré was born in 1854 in Nancy; his father was a physician and professor of medicine at the local university with a family background of distinguished government service. Henri's first cousin Raymond would serve as prime minister of the Third Republic and as president of France during the First World War. Tutored before elementary school by his mother, Poincaré's mathematical ability was recognized early, along with his excellence in written composition (he won first prizes in the *concours générale,* the competition for students of all French lycées).

After attending the École Polytechnique and the École National

Supérieure des Mines, he worked briefly as an engineer and received his doctorate in 1879. Appointed as an instructor in mathematical analysis at the Univerity of Caen, he moved to the University of Paris in 1881, where he remained for the rest of his life as a professor of mathematical physics, mathematical astronomy, and celestial mechanics. He died in Paris in 1912.

As the dominant figure in mathematics at the turn of the century, Poincaré made profound contributions to practically all areas of mathematics as well as to mathematical physics, and his prolific writings exerted a long-lasting influence on the philosophy of science. Philosophically, he was a conventionalist—mathematics and parts of science were, for him, human conventions, which is to say, human *in*ventions. No Platonic universe of ideas for him! Newtonian mechanics was the area in which his work in physics had the most enduring effect, and two ideas in particular are of interest.

The first was his proof of what became known as Poincaré's recurrence theorem: every mechanical system confined to a finite spatial region must necessarily eventually return to any given area, no matter how small, near its starting point. The theorem does not say the system must return precisely to its starting point; as Nicole Oresme had already recognized in the fourteenth century, the probability for that would be nil. By the end of the nineteenth century, however, mathematics was able to deal more satisfactorily with what for the fourteenth century was an insurmountable obstacle. Although Oresme had been correct in denying exact recurrence for almost all systems, there was a recurrence of a fuzzier kind nevertheless.

Poincaré's second contribution of special interest was the discovery of the phenomenon of chaos, which demolished the grand deterministic vision so eloquently expressed by Laplace. The origin of his discovery goes back to his early work in celestial mechanics.[2] He had written a brilliant paper on the motion of several objects, such as planets, influencing one another gravitationally—a notoriously difficult problem when the number is three or greater. However, after it had been published and won a prize from the king of Sweden, he realized to his horror he had made an important error invalidating his conclusion that the motion, as expected, would be quite smooth and predictable. So he spent all his prize money buying up as many of the copies as he could find of the journal in which the erroneous paper had been published, to prevent it from being disseminated. In correcting the error, he found that in most instances the motion of these objects would, in fact, be chaotic, as we would say nowadays.

What prevents the motion of most bodies in interaction with each other from being predictable in any ordinary sense of that word—thus, chaotic—is a phenomenon called sensitivity to initial conditions, as Poincaré discovered. This means that if either their positions or their velocities at the initial instant are altered by small amounts, their speeds and locations at a later time may be vastly different. Two identical systems starting out almost the same way will, after a while, be in entirely different states and no longer resemble one another at all. In other words, even though the Newtonian equations of motion determine their later configuration exactly if their initial data are precisely given—in that sense the laws are deterministic—the slightest error at the start produces a large and unpredictable deviation at a later time, making the motion, for practical purposes, unpredictable.

From a modern perspective, remember that every computer does numerical calculations with a finite number of digits, and all its data necessarily have small errors whose size is determined by the number of digits used. If such a computer is employed to calculate motion, these errors will grow out of control, rendering the calculation eventually useless. This is one reason why weather prediction is so difficult. The grand illusion of having finally attained the long-standing goal of making the workings of nature completely predictable was thus decidedly dimmed by the monkey wrench Poincaré threw into the delicate machinery of classical mechanics in the form of the so-called butterfly effect—a butterfly batting its wing in the Amazon rain forest may cause a tornado in Wisconsin.

We should recognize, however, another important mathematician whose work made a significant contribution not only to Newtonian physics but also to the new physics, exerting a decisive influence on our view of the way nature works. Born in Erlangen, Germany, in 1882, the oldest child of Max Noether, a well-known mathematician, Amalie Emmy Noether was prevented for two years from continuing her education past high school by a rule barring women from full status as university students. After informally listening to lectures on linguistics and mathematics at the University of Erlangen from 1900 to 1902, she managed to matriculate and received her doctorate in 1907. However, since a university career was not open to her, she had to do her research independently. Owing to the constant, strong support of David Hilbert, she was invited to lecture at the University of Göttingen from 1915 to 1916, given unofficial professorial status in 1919, and finally officially made an associate professor in 1922 with a minimal salary. She spent the year 1928–1929 as a visiting professor at Moscow University. Her position in Göttingen, however, lasted only until 1933 when, dismissed as a Jew, she emigrated to the United States, where she became a professor of mathematics at Bryn Mawr College in Pennsylvania until her death of a postsurgical infection in 1935.

While most of Emmy Noether's work was in algebra, her importance in physics rests on a specific theorem that she proved in the context of classical physics but whose validity turned out to be far wider. Noether's theorem states that whenever the equations of motion of any system are invariant with respect to some change in parameters, the system has a corresponding conserved quantity. For example, if a physical system is invariant under translation, that is, it does not change when shifted in space, it is subject to the law of conservation of momentum; if it is invariant under translations in time, that is, it will behave tomorrow as it does today, then its energy is conserved. In other words, Noether's theorem establishes a quite remarkable connection between the symmetries of a physical system and its conserved quantities, a connection that has acquired particular importance for the entire body of modern physics because of the emphasis all contemporary theories put on both symmetries and conservation laws.

Before turning to the upheaval in physics that occurred during the nineteenth and twentieth centuries—related to but not caused by Poincaré's finding of chaos—it is important to describe the great new advances achieved in the areas of sound, light, electricity, magnetism, and heat, as well as those achieved in the final confirmation of atomism. At the end of the eighteenth century, only a small fraction of the phenomena now familiar to physics was known, and of those known, few were understood. Gravity was the only force in nature, other than Hooke's spring force, that had been described mathematically, which limited the application of Newton's laws of motion almost entirely to celestial mechanics. Much remained for physicists to discover, to explain, and to describe mathematically before Laplace's dream of determinism could be realized.

Some of the new physics would lead to basic revisions of Newton's laws, but astonishing as they were, these revisions were not really revolutionary. It was the common phenomenon of heat, as we shall see in the next chapter, that would turn out to contain the seeds of an upheaval spelling the end of the deterministic dream and ushering in a new era in which nature was seen as governed by chance rather than necessity.

six

New Physics

First proposed in the fifth century BCE by Leucippos, subsequently adumbrated by Democritus, and later independently reinvented by the Hindus, the notion that the world is made up of unchanging and indestructible atoms had been resurfacing in science for well over two thousand years. But the idea that underlying the appearance of the great variety of color, sound, shape, and temperature of all we see is nothing but different arrangements of certain colorless, cold, and permanent building blocks did not advance beyond speculation until early in the nineteenth century. The solid scientific evidence for this view got its initial support from John Dalton.

The third of six children of a weaver and devout Quaker, Dalton was born in 1766 in the village of Eaglesfield, Cumberland, England. After attending a Quaker school, he learned mathematics and science largely from public lectures given in nearby Kendal and from the blind natural philosopher John Gough (known from Wordsworth's *Excursion*). He was excluded from Oxford and Cambridge, which at that time were open only to members of the Church of England. Before long, Dalton gave public lectures himself, mostly about meteorology and color blindness—which afflicted both him and his brother—at the Manchester Literary and Philosophical Society, of which he later became president.

In 1793 Dalton was appointed professor of natural philosophy and mathematics at the New College in Manchester, a post he quit after seven years to set up his own Mathematical Academy, offering courses in mathematics, natural philosophy, and chemistry. Though the quintessential outsider—he never married—he was elected in 1816 as a corresponding member of the French Academy of Sciences and in 1822, over his objections, to the Royal Society. He received the Royal Medal in 1826, honorary degrees from the universities of Oxford and Edinburgh, and a government pension in 1833. He died in Manchester in 1844.

Dalton's interest in meteorology led him to study gases, and in 1803 he proposed his law of partial pressures, known ever since as Dalton's law, which states that the total pressure of a mixture of gases in a given volume is the sum of the pressures that each of the constituents would separately exert if they alone occupied the same volume. The explanation he offered for this was that the constituent gases were made up of different particles, each of which exerted repulsive forces upon the particles of their own kind but ignored those of the other kinds. (He had concluded that gases and liquids are made up of particles because gases can be absorbed by liquids such as water, which he explained by postulating that the gas particles occupied interstices between those of the water.)

The following year he announced his chemical law of multiple proportions. A chemical compound is made up of a fixed ratio of weights of its constituents—the law of definite proportions, already known. But Dalton's more informative law of multiple proportions states that if two given elements can combine with each other in more than one compound, their weight ratios in the various cases differ by factors that are small whole numbers. For example, nitrogen and oxygen are able to form five different compounds; in these five oxides 14 grams of nitrogen combine with 8, 16, 24, 32, and 40 grams of oxygen, which differ by factors of 2, 3, 4, and 5. From these two laws he drew the conclusion that the elements must be made up of particles—atoms—which, for a given element, are all alike and have the same weight, and which in chemical compounds attach themselves to one another to form what we now call molecules. In the example of nitrogen oxides, the molecules would consist of one atom of nitrogen and either 1, 2, 3, 4, or 5 atoms of oxygen. From the law of definite proportions he concluded that the weight ratio of elements A and B in a given unique compound of the two must be identical to the weight ratio of their respective atoms. (He was still somewhat confused about the difference between atoms and molecules of gases made up of elements.)

Dalton did not get the constitution of all the molecules right; he thought, for example, that a water molecule consisted of one atom of oxygen and one atom of hydrogen. And his relative atomic weights, based on the assumption that the molecule of the unique compound of A and B must consist of one atom of each, were often wrong. But his basic concept was certainly correct. Atomism had finally been put on a scientific basis, whose solidity was cemented by a great variety of evidence steadily accumulated over the next hundred years. Most of the new physics developed during the nineteenth century made use of it in one way or another, despite a few skeptics such as the respected Austrian physicist Ernst Mach. Today the concept of atoms is regarded as a foundation stone of physics and chemistry. Perhaps the visually most convincing proof was eventually provided after 1827, when the Scottish botanist Robert Brown observed, under his microscope, a mysterious irregular motion of pollen grains suspended in water. These visible movements, thereafter called Brownian motion, were direct evidence for the particulate nature of water, though that fact was not recognized until after 1905, when Albert Einstein explained their cause to be the result of fast-moving water molecules, too small to be seen, colliding with the visible grains.

A refinement of Dalton's discovery was soon provided by the Italian physicist Avogadro, using evidence provided by Gay-Lussac. In 1802 the French chemist Joseph Louis Gay-Lussac (1778–1850) had announced a universal law governing the expansion of gases, namely, that for a given rise in temperature all gases expanded by the same fraction of their volume. The French physicist Jacques Alexandre César Charles (1746–1823) had already discovered this fact in 1787 but had not published it; for this reason it is known today as Charles's law. Furthermore, in 1808 Gay-Lussac found what is now called Gay-Lussac's law of combining volumes: when two gases combine chemically, the volumes they and the product (if a gas) occupy at a fixed pressure always stand in simple numerical ratios. For example, two gallons of hydrogen and one gallon of oxygen yield two gallons of steam.

Amedeo Avogadro was born in Turin in 1776, son of a distinguished lawyer. He too started out as a lawyer and did not take lessons in mathematics and physics until the age of 24, but he made rapid progress. In 1820 he was appointed to the newly established professorship in mathematical physics in Turin, the first such professorship in Italy, which temporarily fell victim to the political turmoil at the time before being re-established permanently and held for a short while by Cauchy; there Avogadro remained until his retirement in 1850. Subsequently regarded as the founder of physical chemistry, he was a man of great modesty, who never received any recognition during his lifetime for his important work. He died in Turin in 1856.

Avogadro's most significant achievement was the formulation of Avogadro's law, an idea that Dalton had already considered but unwisely rejected. It states that at any given temperature all gases contain the same number of molecules per unit volume. Basing his hypothesis on Charles's law of gas expansion developed by Gay-Lussac (who also invented the word *molecule*), Avogadro then used it to explain Gay-Lussac's law of combining volumes in a rational fashion, thereby lending his hypothesis additional support. Considering the large differences in weight, and therefore presumably in size, between molecules of various gases, Avogadro's law would be rather surprising if these molecules were closely packed, as Dalton believed. His law therefore lent support to the view that there are relatively large spaces between the molecules of a gas. Despite its ample verification in many contexts, this law is still often referred to as Avogadro's hypothesis.

Note that Avogadro's law can be used to find the relative weight—relative to that of hydrogen—of each molecule of a gas directly from the density of that gas: the ratio of the weight of each molecule of gas A to that of each molecule of hydrogen is equal to the ratio of the weights of a unit volume of A to that of a unit volume of hydrogen, that is, the ratio of the density of A to the density of hydrogen. This was a more reliable method of determining molecular weights than Dalton's. By translating Avogadro's law into weights rather than volumes, we state it the following way, which avoids the need to mention temperature: the number of molecules in a gram mole (a quantity of gas whose weight in grams is equal to its relative molecular weight) of any gas is a universal constant, now known as Avogadro's number (whose numerical value, 6.022×10^{23}, was not established until the end of the nineteenth century). In other words, this is the number of molecules in 2 grams of hydrogen or in 32 grams of oxygen or in 18 grams of water. Each molecule of hydrogen gas contains two hydrogen atoms, each molecule of oxygen gas contains two oxygen atoms, and each molecule of water contains two hydrogen atoms and one oxygen atom. That is why two gallons of hydrogen and one gallon of oxygen make two gallons of steam. Chemists refer to the number $N = 6.022 \times 10^{23}$ of molecules required to make a gram mole of gas as a *mole.*

One area of physics that reached theoretical completion on the basis of the particulate nature of gases, together with Newtonian mechanics, was acoustics. Until the middle of the seventeenth century the dominant view was that the sound of ringing was produced by a stream of particles originating from the bell. Galileo, as well as his younger contemporary, the French natural philosopher, theologian, and mathematician Marin Mersenne (1588–1648), found that the pitch of a musical note is proportional to the frequency of the vibration producing it. Prizing precision in scientific work above all else, Mersenne also discovered that the intensity of sound decreased as the square of the distance from its source, and that its speed was independent of its pitch and loudness. Today, he is regarded as the founder of the field of musical acoustics.

The myth that sound consisted of a stream of particles, and therefore required no air to be heard, was finally demolished by an experiment still performed in physics classes to this day, first done by the German Jesuit scholar Athanasius Kircher (1601–1680) and described in his book *Musurgia universalis,* published in 1650. The demonstration consists of a ringing bell enclosed in a tightly closed jar, from which the air is gradually removed by means of a pump. The ringing we hear gets dimmer and dimmer until it finally becomes inaudible when the jar is totally evacuated, thereby showing convincingly that air was required for the propagation of sound. Unfortunately, Kircher's apparatus was unable to achieve a good enough vacuum to extinguish the sound completely, and he erroneously concluded that air was unnecessary for the propagation of sound. It took the much improved air pump devised by Robert Boyle to make Kircher's experiment really persuasive.

This is where knowledge of sound stood at the time of Newton, whose equations of motion would finally make a detailed understanding of the propagation of these vibrations in the air possible. The first mathematician to employ Newton's equations for a detailed study of vibratory motion was Daniel Bernoulli. During the nineteenth century, the different characteristics of sound in various gases were studied by the Irish physicist George Gabriel Stokes (1819–1903), who made many other contributions to the field of fluid dynamics as well. He also suggested that under certain circumstances sound waves would undergo sharp discontinuities, a phenomenon we now call shock waves; but he was persuaded to retract this proposal because it appeared to contradict the conservation of energy. Investigating the physiology of hearing, the German physicist Hermann von Helmholtz developed a theory of musical timbre in which he recognized that the variations in the timbre of the same notes produced by various musical instruments are caused by the admixture of different overtones. Finally, the field of acoustics was brought to its theoretical completion by the great English physicist John William Strutt, better known as Lord Rayleigh (though the practical application of acoustics in the construction of concert halls still bedevils architects to this day).

John William Strutt, born in 1842 at Langford Grove in Essex, England, was the son of Baron Rayleigh, whose title he inherited upon his father's death. As a boy, he was forced by poor health to withdraw from Eton and Harrow and be privately tutored, but at the age of eighteen he entered Trinity College, Cambridge, from which he graduated in 1865, becoming a Fellow the following year. He devoted the rest of his life to the pursuit of science, an uncommon step for a member of the landed aristocracy, who usually made their careers either in the Church or the military if they were not content to live the life of a country squire. Instead of taking the customary Grand Tour of the European continent, Strutt chose to visit the United States for a couple of years, and after his return he set up his own research laboratory at the family estate, Terling Place, where he performed most of his subsequent work in physics. In 1871 he married Evelyn Balfour, sister of Arthur James Balfour, the man who would later become prime minister and serve as foreign secretary during the First World War.

From 1879 to 1884, Lord Rayleigh, as he was now titled, abandoned his laboratory to become Cavendish Professor of Experimental Physics at Cambridge University, succeeding James Clerk Maxwell, who had died unexpectedly early. Elected to the Royal Society, which he would later serve as secretary, he accepted the position of president of the British Association for the Advancement of Science and made another visit to America before resigning the Cavendish professorship and returning to Terling, which remained his primary residence and scientific headquarters from then on. In 1908 Rayleigh was appointed chancellor of Cambridge University, an honorary post he retained until his death at Terling in 1919.

One of the most productive theorists and experimenters of the nineteenth century, Lord Rayleigh made fundamental contributions to all areas of what is now called "classical" physics, as well as sowing some of the seeds of the revolution to come (for which he had no taste). His discovery of the element argon simultaneously with the chemist William Ramsey earned him the 1904 Nobel Prize in physics, while Ramsey received the Nobel in chemistry. (The annual ritual of Nobel Prize awards had begun in 1901 and quickly became, in the minds of many people, the touchstone of scientific achievement.) Most of his theoretical work dealt, in one form or another, with wave propagation, a subject that crossed many boundaries—his theory of the scattering of light by small particles, still called Rayleigh scattering, led to an explanation of why the sky is blue—and he made his greatest contribution, and the one most relevant to our narrative, in his two-volume treatise *The Theory of Sound.*

The writing of this work had begun during an extended journey by houseboat up the Nile, together with his young wife, in order to recuperate from a serious attack of rheumatic fever. *The Theory of Sound* contained detailed mathematical descriptions of the nature of sound as a longitudinal wave made up of minute variations in the pressure of its surrounding fluid or solid, caused by differences in the density of the constituent particles—alternating condensations and rarefactions—and its propagation in different media and under various conditions. These pressure variations originate from vibrations in its source, be it a musical instrument, a bell, or an explosion, and they travel at a fixed velocity that varies with the medium but is independent of the velocity of the source. However, a moving source vibrating at a given frequency produces sound of a higher frequency in its forward direction and of a lower frequency in its backward direction because the pressure waves are compressed in one direction and stretched out in the other. This effect, audible as a drop in pitch when a moving sound source such as a train passes a listener, was discovered in 1842 by the Austrian physicist Johann Christian Doppler (1803–1853), after whom it was named. In its many editions, which were kept up to date on such evolving developments as binaural effects in human hearing and newer ideas on shock waves (the subject that Stokes had abandoned), Rayleigh's treatise remains the bible on sound propagation to this day.

The nature of the other primary sensory medium of human perception, light, also remained for the nineteenth century to untangle. At the end of the eighteenth century the rivalry between Newton's corpuscular theory and Huygens's wave theory was unresolved. The decisive experimental blow against Newton was delivered at the very beginning of the new century when the English physicist Thomas Young (1773–1829) discovered the phenomenon of interference. In addition to being a physicist, Young, a Quaker, was a physician and an accomplished linguist, proficient not only in Greek and Latin but conversant with eight Middle Eastern languages, from Hebrew to Ethiopic, and among the first to decipher Egyptian hieroglyphics. In 1801 he performed one of the truly crucial experiments in the history of physics. His arrangement consisted of a beam of monochromatic light shining first through a slit in a screen, then falling on two narrow parallel slits on a second screen, and finally shining on a third screen, where it was observed (Fig. 10).

To the consternation of supporters of the particle theory, the image on the third screen did not show two bright stripes in a field of darkness, as would be produced by a stream of particles, but several bands of alternating light and dark stripes, generally referred to as "fringes" (Fig. 11). Young's explanation of this curious phenomenon was that the fringes were produced by interference—light waves from the two slits successively adding or subtracting one another, depending on the difference in the distance of a given spot on the third screen from the two slits in the second—thereby conclusively showing that light was a wave. He thought of it as analogous to a sound wave; in this, however, he was to be proven wrong within twenty years.

Young's diffraction experiment was independently duplicated in 1815 by the French physicist Augustin Jean Fresnel (1788–1827). Unaware of the work of either Huygens or Young and experimenting in his spare time only (he had a government position as a civil engineer), Fresnel specialized in research on the properties of light with the goal of disproving the particle theory and demonstrating its wave nature. At the suggestion of Poisson (whose intent was to disprove the wave theory), he also performed an experiment of directing a beam of light at an opaque circular disk and, to his delight and Poisson's dismay, he found a bright central spot in the middle of the shadow on a screen behind it: another result of interference and proof of the wave theory, though it is not entirely clear that Poisson was convinced. Fresnel then went on to construct a detailed mathe-

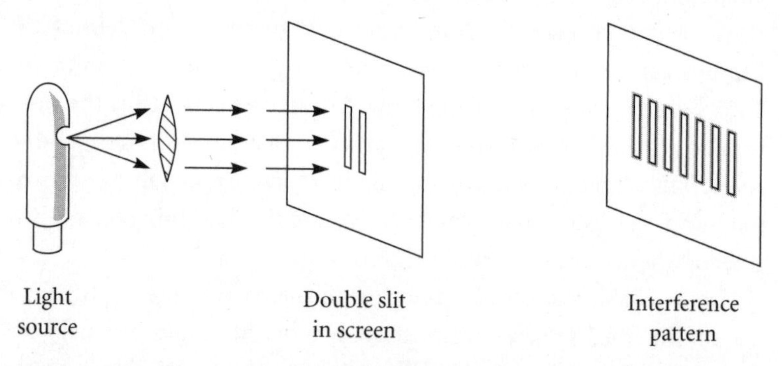

Figure 10 Interference in a double-slit experiment, the way it might be demonstrated now rather than as Young did it. (From Fig. 18.41(a), p. 926, K. W. Ford, *Classical and Modern Physics*, vol. 2, Xerox College Publishing, 1972.)

matical description of the interference phenomenon in more general contexts.

In the meantime, the French physicist Étienne Louis Malus (1775–1812) had discovered that, under certain conditions, light possessed a property he called polarization. This conclusion was based on his analysis of experiments with light shining through birefringent crystals such as calcite, which split a ray shining on its surface into two rays refracted at different angles and with different properties, as can be demonstrated by allowing these two to shine on a second piece of the crystal: the two rays are differently polarized. (The polarization of light can be demonstrated more simply today with Polaroid sunglasses.) Fresnel's explanation was that light must be a transverse wave, with the vibration of a polarized beam swinging at right angles to the direction of its propagation, rather than a longitudinal wave such as sound, in which the pressure vibrations oscillate along the line of propagation. But a wave of what? Sound was a wave of vibrations of a medium such as air, but light required no air or other gas;

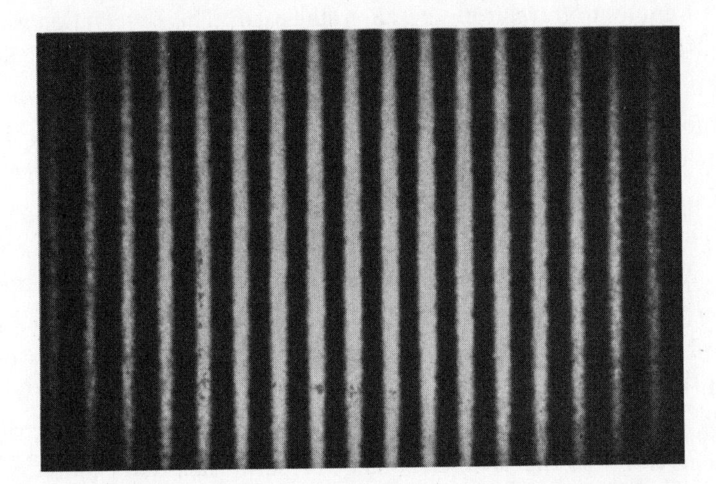

Figure 11 Interference fringes produced by two slits. (Brian J. Thompson, Institute of Optics, University of Rochester. From Fig. 18.41(b), p. 926, K. W. Ford, ibid.)

it went right through a vacuum. Some other kind of medium must be carrying it, and the name chosen for this mysterious phenomenon, the nature of which remained to be unraveled, was *ether*, echoing Aristotle.

The velocity of light had been measured for the first time in 1679 by the Danish astronomer Ole Christensen Römer (1644–1710), based on variations in the times at which the planet Jupiter eclipsed its moon Io, depending on Jupiter's distance from the earth. His successful prediction of a delay by ten minutes of the next expected eclipse, and his explanation of the reason for this delay, caused a sensation. The notion that light traveled with infinite speed was still accepted by many, including Descartes. The speed implied by Römer's measurements, 225,000 km/sec (140,000 miles/sec), was remarkably close to the now-accepted value of 299,792 km/sec and so large that its measurement in a laboratory on earth would present enormous difficulties. Römer did not actually calculate the speed but only stated that light would take about 22 minutes to cross the earth's orbit, and 11 minutes to travel from the sun to the earth. This observation was subsequently converted into a numerical velocity by Huygens. The feat of measuring the speed of light in a laboratory was finally accomplished by the French physicist Armand Hippolyte Louis Fizeau (1819–1896), who also found that light traveled more slowly in water than in air, another nail in the coffin of particulate theories of light: Newton had predicted that it should move faster in water.

Along with light, two other puzzling phenomena had been under scrutiny in the seventeenth and eighteenth centuries: electricity and magnetism. Both had been known since antiquity. Thales of Miletus is credited with the discovery that a piece of amber, after being rubbed, will attract other small objects to itself; the word electric stems from the Greek word for amber. And the fact that a lodestone attracted iron was mentioned by Lucretius. The magnetic compass was used for navigation by the Chinese and by Western Europeans in the twelfth century, by the Arabs in the thirteenth, and by Scandinavians in the fourteenth century. In the sixteenth century William Gilbert tried to understand lodestones and compasses (though not amber), and his ideas, collected in his treatise *De magnete*, inspired Kepler in his search for an explanation of the motion of the planets.

During the seventeenth century, much of the pursuit of mathematics and physics on the continent was in the hands of Jesuit scholars and teachers. Whenever a conflict between science and religion arose, they resolved it by regarding scientific results as mere mathematical models rather than the truth. When no conflict appeared, some Jesuits used science to defend their theology. The most important of these was the widely admired Athanasius Kircher, who, after moving from Germany to Rome because of the Thirty Years' War, assembled a circle around him there. Much baffled by strange influences on electrostatic phenomena exerted by ambient humidity, of which they were unaware, they argued at length about whether electric and magnetic attractions were caused by effluvia and currents in the air surrounding electrified objects. The principal properties of electric and magnetic bodies were, they assumed, a certain fattiness and glueyness. Descartes, on the other hand, constructed a theory of "electrics" that, though simplistic, did away with any effluvia and stickiness, in favor of corpuscles and vortices.

In England, interest in electrics and magnetics was kept alive after Gilbert by the Jesuit theologian Thomas White (1593–1676), better known as Blacklow, and a few others. Experiments by Robert Boyle using his vacuum pump did much to elevate the scientific discourse, discrediting all theories involving the air in electrical attraction. Otto von Guericke contributed to the discussions as well, particularly because of his demonstration of electric repulsion as well as attraction. Huygens also became interested in electric phenomena, performing experiments and applying Cartesian ideas to Guericke's discovery of electrostatic repulsion, though he published none of these results.

Shortly after his installation as chief experimentalist of the Royal Society, which coincided with the tenure of Isaac Newton as president, Francis Hauksbee (c. 1666–1713) publicly demonstrated a new electrostatic effect. Employing an "electrical machine" to perform elaborate demonstrations of electrostatic repulsions, he astonished his audience by rubbing the exterior of a partially evacuated glass globe, thereby producing an intense glow of light within it, which gradually disappeared as air was allowed back into the globe. Another new electric phenomenon was discovered by Stephen Gray (1666–1736), an inspired amateur who managed to become a member of the Charterhouse foundation and devoted most of his time to electrical experimentation. Electric attractions, he found, could be communicated from one object to another through the air—sometimes with a visible spark—and through other bodies, a property that he investigated and publicly demonstrated in a great variety of ways.

The capstone of eighteenth-century discoveries of electrostatics was provided by the French infantry officer Charles François de Cisternay Dufay (1698–1739), who, immediately upon his retirement at the age of 25, became a member of the Paris Academy of Science (assisted by family connections). He found not only that all nonmetal objects solid enough to rub could be electrified and that metals communicated "electrical virtues" differently from other substances, but also that two different kinds of electricity existed, accounting for attractions and repulsions. Some substances always generated one kind, others always the second kind. Objects with unlike electrification attracted one another, while those with like electrification repelled one another.

By about 1740, mostly stimulated by Gray and Dufay, electricity had become a very fashionable topic of intellectual and genteel discourse. This is when Benjamin Franklin (1706–1790) entered the picture. Going much beyond the observation of electric attractions and repulsions, Franklin experimented with Musschenbroeck's magic bottle, a device invented in the 1740s by Pieter van Musschenbroeck (1692–1761) of Leiden that became known as a Leiden jar. It facilitated the accumulation of large amounts of static electricity, and Franklin fascinated visitors to his laboratory by using these jars to produce long sparks from sharply pointed metallic objects. His famous experiment with an iron key dangling from a kite in a thunderstorm demonstrated that lightning was an electric phenomenon. Franklin's most famous practical application of this insight, together with his experiments using pointed metal objects, was his invention of the lightning rod.

The English natural philosopher Henry Cavendish (1731–1810), who discovered hydrogen and was the first to measure Newton's gravitational constant directly on earth (by measuring the minute attraction between two lead spheres), proposed a single-fluid theory for the flow of electricity. He also found that the electric attraction between two oppositely charged objects varies as the inverse square of the distance between them, a law that had been first conjectured in 1767 by the English chemist and Presbyterian theologian Joseph Priestley (1733–1804), the reluctant discoverer of oxygen. Drawing on Priestly's work and that of others, Cavendish intended to write a great treatise on electricity that would be analogous to Newton's *Principia.* But he never did, and owing to his extremely reclusive—by some accounts, almost autistic—nature, most of his work remained unknown for some fifty years. Today the physics laboratory at Cambridge University is named in his honor.

Priestley was an outspoken dissenter from the Church of England and a vocal supporter of the French Revolution, which made him extremely unpopular in Birmingham, where he lived. When even a move to London did not alleviate the hostility surrounding him, he emigrated to America. Rejecting an offer of a professorship at the University of Pennsylvania, he settled in Northumberland, Pennsylvania, where he died in 1804. Priestley had been a promoter of the phlogiston theory, which held that combustion produced the gas phlogiston in air. But then he discovered that, on the contrary, combustion as well as respiration required the presence of a gas that he called "dephlogistinated air." It was Lavoisier, a strong opponent of the phlogiston theory, who named it oxygen.

The French physicist Charles Coulomb (1736–1806) is credited with the discovery that the force between two electrically charged objects is proportional to the product of the charges they carry and inversely proportional to the square of the distance between them (provided that the size of these objects is small compared to their distance), and similarly for the magnetic force. The instrument both he and Cavendish employed for measuring these tiny forces was a torsion balance, whose degree of twist could be sensitively measured. Coulomb is commemorated by having the inverse-square law of electrical force as well as the unit of electrical charge named after him. Poisson later formulated and generalized Coulomb's experimental results in mathematical terms, transferring them to magnetism as well. He also invented two-fluid theories for electricity and for magnetism, both of which were later discarded.

While Coulomb and Cavendish dealt exclusively with static electricity, Franklin also experimented with the flow of electric currents through wires. For the source of these currents he employed the voltaic pile, which had been invented in 1800 by the Italian physicist Alessandro Volta (1745–1827). A forerunner of the modern battery, this invention, which produced strong electric current, consisted of alternating layers of silver and zinc discs separated by brine-moistened cardboards. Its use led to the important discovery by the German physicist Georg Simon Ohm (1789–1854) that the current through a wire varied in inverse proportion of the wire's resistance, a quantity that is proportional to its length and inversely proportional to the area of its cross section (depending also on the material the wire is made of).

A discovery that the Danish physicist Hans Christian Oersted (1777–1851) made in 1820, however, would have much more far-reaching consequences. As he had anticipated on philosophical grounds—in his opinion all forces ought to be interrelated—he found that a wire carrying an electric current produced magnetism: it could deflect a compass needle. He had been looking for such an effect for seven years without finding it because, whereas he had expected the magnetic force to be in the same direction as the current, it turned out to be perpendicular to it. Thus electromagnetism was born; the two phenomena of electricity and magnetism were intimately connected.

This connection was further strengthened when the French mathematician, physicist, and chemist André Marie Ampère (1775–1836), stimulated by witnessing a demonstration of Oersted's discovery, found that two parallel current-carrying wires attracted or repelled one another, depending on whether the currents in them ran in the same or in opposite directions. The conclusion had to be that not only did electrical currents produce magnetic forces, but these magnetic forces in turn also acted on currents. He presented all the results of his studies in mathematical form in his *Mémoire sur la théorie mathématique des phénomènes électrodynamiques uniquement déduite de l'expérience,* and the mathematical law embodying the relation between electric currents and magnetic force now carries his name, as does the unit of current.

If Oersted and Ampère discovered that electric currents produce magnetism, the inverse effect, that magnetism can produce currents, was found by the American physicist Joseph Henry and the English physicist and chemist Michael Faraday. Henry (1797–1878), the son of a laborer, was born in Albany, New York, and educated at the Albany Academy. After discovering in 1830 what came to be called electromagnetic induction, he was appointed professor of natural philosophy at New Jersey College (subsequently renamed Princeton University) and later became the first president of the National Academy of Sciences of the United States. A year after Henry's discovery, the same induction phenomenon was independently found in England by Faraday, who had been searching for this effect since François Arago (1786–1853) announced in 1824 that a rotating copper disk caused a compass needle above it to be deflected. It happened that Faraday published his discovery before Henry, and the law embodying it came to bear Faraday's name. Apart from tightening the knot between electricity and magnetism, induction turned out to have important practical applications, governing the functioning of transformers as well as electric motors and generators. The first electric generator was built by Faraday himself.

The son of a blacksmith, Michael Faraday was born in 1791 in Newington, Surrey. With little schooling but a voracious appetite for reading the *Encyclopedia Britannica,* he was sent as an apprentice to a book binder in London, but grew up to be a scientist with an extraordinary genius for experimentation. His brilliant promise was discovered and nurtured by the great chemist Humphry Davy (1778–1829), who took him on a tour of Europe, during which Faraday became acquainted with many of the top scientists of the day. On his return to England, he obtained a position at the Royal Institution, where he remained to make most of his discoveries in chemistry and physics. A renowned scientist, he became an extremely popular scientific lecturer to the public as well. However, at the age of 47 he suffered a breakdown from which he never fully recovered. Though he returned to his research six years later, with renewed productivity, he never fully regained his mental powers, possibly as a result of inadvertent poisoning during his earlier chemical work. After resigning his position at the Royal Institution in 1862, he retired to live, courtesy of Queen Victoria, in an apartment at Hampton Court, where he died in 1867.

Faraday made many fundamental contributions to organic chemistry and discovered the basic laws of electrolysis; in physics, his most significant work focused on electricity and magnetism. In addition to the discovery of electromagnetic induction, he introduced an idea that turned out to have important consequences in physics to this day. In order to avoid the repugnant concept of action at a distance that underlay Newton's gravitational force and threatened to raise its ugly head again to account for electric and magnetic forces, he conceived of the notion of a field. Instead of thinking of an electric charge at point A in space directly exerting a force on another charge at a distant point B, the charge at A would produce a condition in space everywhere, called a field, consisting of "lines of force" that would act on the corresponding charge at point B directly. Henceforth electromagnetic influences would always be described in terms of electric and magnetic fields. However, he lacked the mathematical tools to flesh out this idea, a task subsequently accomplished by Maxwell.

Faraday also made the discovery that the polarization of a ray of light could be changed by a magnetic field, an experimental observation which clearly indicated that the nature of light was related to electromagnetism. Exploiting this indication, shortly after his recovery from his nervous breakdown, Faraday put forward the idea in 1846 that light was simply a rapid oscillation of the electromagnetic field, explicitly rejecting the need for a medium such as ether to carry the vibrations. A bold thought indeed, jumping ahead of Maxwell to Einstein! Further evidence for the relation between light and electromagnetism was provided by the fact that the precise mathematical expressions for Ampère's and Faraday's laws both contained within them as a numerical factor the ratio of the different units of electric charge employed in electrostatic and electrodynamic measurements, such as dealing with moving charges in currents. The two German physicists Wilhelm Eduard Weber (1804–1891) and Rudolph Herrmann Arndt Kohlrausch (1809–1858) discovered in 1855 that the ratio of these units was equal to the speed of light, though they apparently did not regard their discovery as especially significant.

Born in Edinburgh, Scotland, in 1831, the son of a lawyer, James Clerk Maxwell grew up as the only child in a well-to-do, intellectually oriented family. His mother died when he was eight years old, and a year later an aunt took him from the charge of an incompetent tutor, who considered him a slow learner, and enrolled him in the

Edinburgh Academy. Fascinated with mechanical models and geometry, he published his first scientific paper at the age of fourteen. Two years later he entered the University of Edinburgh, remaining there for only three years and moving on to Cambridge University in 1850. There, his mathematics tutor recognized his exceptional powers in any area of physics but bemoaned his lesser talent in mathematical analysis. He declined a fellowship in Trinity College and, because of his father's failing health, returned home. Soon he was appointed professor of natural philosophy at Marischal College in Aberdeen and married the daughter of its principal two years later. When Marischal College and King's College, Aberdeen, merged in 1860 to form the new University of Aberdeen, Maxwell was appointed to the professorship of natural philosophy at King's College, London, and one year after that he was elected to the Royal Society.

In 1865, after his father's death, Maxwell resigned his professorship and retired to his family estate, Glenair, in order to devote himself to the research needed to complete his monumental *Treatise on Electricity and Magnetism.* Returning to Cambridge University in 1871 as its first professor of experimental physics, he set up the Cavendish Laboratory in 1874. In 1879 Maxwell died of the same kind of cancer as his mother, and at the same age.

Initially working on color vision, in the course of which he produced the first color photograph (employing a three-color process), Maxwell next attacked the problem of Saturn's rings, whose stability had long resisted explanation. He showed that, if a ring was made up of many small orbiting objects, it could be stable. But the achievement on which his fame rests more than any other took him many years to accomplish, beginning in 1855. Stimulated by Faraday's ideas, he planned to construct a complete theory of the electric and magnetic fields, including light as one of their manifestations. Inveterate model builder that he was, he constructed the entire impressive mathematical edifice on the basis of an elaborate mechanical model of the structure of the ether. The mathematical theory has remained (and has served for the calculation and explanation of electromagnetic effects and as a standard model for many theories describing other fields to this day), but the underlying mechanical construct was discarded almost immediately. As Heinrich Hertz put it in his own book *Electric Waves,* "Maxwell's theory is Maxwell's system of equations."

That system of differential equations (constructed by means of the differential calculus) allowed the calculation of all electric and magnetic fields—always appearing in combination—caused by electric charges and by magnets, whether moving or at rest, as well as the resulting forces acting on such charges and magnets. (The forces exerted by electromagnetic fields on moving electric charges were explicitly determined from Maxwell's equations by the Dutch physicist Hendrik Antoon Lorentz [1853–1928], and they bear Lorentz's name to this day.) His theory explained the nature of light as a combination of rapidly transversely oscillating electromagnetic waves, the perceived color depending on its wavelength. At the same time, the equations predicted the existence of waves of the same kind but of longer wavelengths to which the human eye is insensitive. In 1888, such waves, later called radio waves, were discovered by the German physicist Heinrich Hertz (1857–1894), whose name now serves as the unit of frequency: one hertz means one oscillation per second.

Maxwell's second great contribution to our understanding of nature was his kinetic theory of gases, to which we shall return in Chapter 8. His electromagnetic theory brought to a close, in the language and with the tools of what we now call classical physics, the puzzles posed by light, electricity, and magnetism. On the other hand, his kinetic theory of gases, as we shall see, marks the beginning of a new paradigm, in which probability takes the place of deterministic prediction. He thus straddled the upheaval that shook physics during the nineteenth and twentieth centuries. But before exploring that story, first we must take up the last branch of physics left over from the eighteenth century and completed in the nineteenth, dealing with heat.

Empedocles as well as Aristotle regarded fire (by which they meant heat) as one of the elements making up the world. That a hot body warmed up a cooler one in contact with it was known, of course, but the nature of the heat that seemed to flow from one to the other in this process remained a mystery even in the eighteenth century. Stimulated by his discovery of specific heat (the amount of heat required to raise the temperature of one gram of a substance by one degree), the Scottish chemist and physicist Joseph Black (1728–1799) in 1760 reintroduced the idea, going back to Lucretius and Heraclitus, that heat was a substance: an indestructible fluid that filled the interstices of all materials and that flowed from a body of higher temperature to one of lower temperature in contact with it, as naturally as water flows downhill. (Galileo is usually credited with inventing the modern method of measuring temperature, though the thermometer was also almost simultaneously invented by Cornelius Drebbel in Holland and the idea of a heat-measuring device had been around since antiquity.) Black's caloric theory, adopted by Lavoisier, held sway until well into the middle of the nineteenth century, though it had to compete all the while with the older *vis viva* or kinetic theory of which Robert Boyle had been a champion and which Daniel Bernoulli had also favored.

The decisive experiment overthrowing the caloric theory, at least in the minds of many scientists, was performed by Count Rumford. This physicist and colorful soldier of fortune was born Benjamin Thompson in 1753 in Woburn, Massachusetts, but later moved to Rumford (now Concord, New Hampshire), where he married a wealthy widow. This match lasted but three years, when they permanently separated. As a loyalist and possibly a secret agent of the Crown, he was forced to flee to London during the American Revolution, returning as an officer with a British regiment. After the war, Thompson lived permanently in Europe, eventually in Bavaria, where he was appointed minister of war and head of the Bavarian army, as well as grand chamberlain of the elector, who made him a count of the Holy Roman Empire. As a count, he adopted the name Rumford.

In the meantime, a lengthy scientific paper he had written on the properties of gun powder and cannon vents led to his election to the Royal Society, and when he returned to England, he became a co-founder (with Joseph Banks) of the Royal Institution, whose purpose was, and still is, the popularization of science. He also endowed a substantial scientific prize that was named the Rumford medal. Finally, he settled in Paris, where he married the widow of Lavoisier, though that match did not last long either. Count Rumford died in 1814 in Auteuil, near Paris, in his will endowing the Rumford professorship at Harvard University.

Always intensely interested in the properties of heat—he designed new uniforms for soldiers to mitigate their loss of body heat and devised both a more efficient domestic range and a better fireplace—Thompson made an observation in 1798 that would have a decisive effect on the controversy concerning the nature of heat. As a military commander in Bavaria, he was concerned with the manufacture of cannons, which were bored from iron blocks by means of drills, a process that made the barrels red hot. According to the caloric theory, the cutting of the drills into the iron allowed heat in the form of caloric fluid to escape. However, he observed that while a very blunt drill bit would cut no iron, it would generate even more heat. The heat had to be produced by the work done in turning the drill, a conclusion which supported the idea that heat was a form of vibratory motion of the constituents of matter. (At the same time he had, in effect, performed the first experiment showing the conversion of mechanical work into heat.) Five years later, Humphry Davy came to the same conclusion after melting two pieces of ice by rubbing them together. If heat could be produced by means of mechanical work, how could it be a fluid flowing through matter? Decisive as the argument seems to us, it took another half century for the caloric theory to disappear entirely.

As a completely phenomenological theory, the newly born science of thermodynamics did not depend on whether the caloric or the kinetic theory carried the day. It dealt with the behavior of heat, its conduction through objects, and its transfer from one body to another, as well as its relation to other phenomena such as pressure and volume changes of gases, rather than with its fundamental nature. Some of the main contributors to thermodynamics continued to believe in the caloric fluid. What stimulated the development of this science in the nineteenth century was primarily the Scottish engineer James Watt's 1765 invention of the condenser for the steam engine, which became the driving force, figuratively and literally, of the Industrial Revolution in Europe and America. While the technical improvement of the steam engine was the job of engineers like Watt, understanding the fundamental principles by which it functioned was the province of physicists, though the two objectives often overlapped.

The first step was to transform Benjamin Thompson's qualitative observation into a quantitative law. This was done by the British physicist James Prescott Joule (1818–1889), who had received some of his early scientific instruction from John Dalton but was otherwise scientifically self-taught. Supporting his laboratory research with his own funds (as the son of a wealthy brewery owner he could afford that) and never taking an academic position, Joule did much of this research in his off-hours before and after his work at the family's brewery. His great strength as an experimenter lay in his inspiration and in his patience and extreme precision in measuring temperatures with unprecedented accuracy.

Using these skills, he determined the mechanical equivalent of heat. First, he arranged a set of paddles rotating in a bucket of water, thereby raising its temperature, which allowed him to measure exactly how much work, turning the paddles, was required to raise the temperature of a measured amount of water by one degree. Similarly, he managed to determine the electrical equivalent of heat by having an electric motor turn the paddles, and he also measured the heat produced in the process of electrical conduction. After the first publication of his result in 1843 was totally ignored, he published an improved version with very high precision in 1845. However, he had difficulties getting his results known before William Thomson (later to become Lord Kelvin) and George Stokes took notice of them in 1847. Nevertheless, Joule established that in all these processes the total amount of energy was conserved. It can be converted from mechanical or electrical form to heat, but none of it will be gained or lost in the process. What is more, the amount of heat obtained from a given quantity of work does not depend on the method used for the conversion or its speed. This justifies the definition of *energy* as a physical concept that manifests itself in different forms in various parts of physics, convertible from one into the other but always conserved. (When your exercise machine tells you the amount of calories you have burned up after a workout, it makes use of the conservation of energy, converting the amount of work you have done into heat, measured in calories.)

At the beginning of the nineteenth century, the notion of energy was not yet well established; physicists used the words *force* and *energy* interchangeably, though *force* also meant the push needed, according to Newton's laws of motion, to accelerate an object. Just as the understanding of what was meant by *caloric* was tied to the idea that this heat-fluid was conserved, so it took the discovery of conservation in the conversion from one form to another to clarify the meaning of energy, a concept that today is regarded as one of the most fundamental in all of physics.

A priority dispute, exacerbated by national pride, somewhat complicated the history of the conservation law. In addition to Joule, a second player in this drama was Robert Mayer, born in Heilbronn,

Germany, in 1814. Mayer studied medicine at the University of Tübingen, where he showed little intellectual promise and irked the authorities with his rebelliousness and stubbornness—for example, he reacted to a one-year suspension from the university with a six-day hunger strike. After passing his doctoral examination, he took a position as ship's surgeon on a vessel sailing to the East Indies, which left him bored and restless but keen-eyed: he noted a distinct change in the color of sailors' venous blood as they traveled from Europe to the tropics and concluded that in hot climates less work was required than in cool surroundings for the extraction of oxygen from the blood in order to keep the body at its normal constant temperature. From this he arrived at his conservation law by a somewhat convoluted chain of reasoning, in part oversimplified and in part simply wrong. However, he recognized that underlying his observation was something fundamental; as chemistry, in his view, was based on the conservation of mass, so physics should be the science bound together by a "conservation of force."

When, upon his return, he wrote up his conclusions, based on an erroneous understanding of mechanics, and submitted the paper to the *Annalen der Physik und Chemie* in 1841, it was rejected. With the assistance of a mathematician he improved his arguments, and the new version, which included a calculation of what amounted to the mechanical equivalent of heat, was published in 1842 in the *Annalen der Chemie und Pharmacie.* This publication included only the result of his calculation, however, without offering any of the needed details; these he supplied in an extended paper in 1845, the very year in which Joule published his own impressively precise experimental result (which differed from Mayer's by more than 15 percent). Mayer's detailed 1845 paper, however, was rejected by the *Annalen der Chemie und Pharmacie,* and he published it privately. Although distributed widely, to his great disappointment it attracted no attention whatsoever.

Mayer was now visited by severe personal problems: the death of three of his children within three years and the deterioration of his marriage. Suffering a serious breakdown, he made an unsuccessful suicide attempt in 1850, entered a sanatorium to recuperate, and was subsequently committed to an asylum. Released in 1853, the doctors having given up on him, he returned to Heilbronn and cautiously resumed his medical practice while, for ten years, avoiding all scientific work. This was an intense period in the development of thermodynamics, when the energy concept became accepted, but Mayer's work remained unappreciated. The scientific world's lack of attention to him went as far as an erroneous report of his death by the editor of the *Annalen der Chemie und Pharmacie.*

In the 1860s Mayer, though scientifically retired, suddenly became the center of a heated priority dispute instigated by the Irish physicist and popular lecturer John Tyndall (1820–1893). While preparing a series of lectures on heat, he was informed by Rudolf Clausius, with support from Hermann von Helmholtz, that Mayer had been the first to understand the concept of energy and its conservation, and Clausius sent him copies of Mayer's papers. Having previously regarded these papers as of no consequence, Clausius seems to have carefully read them for the first time on this occasion. From that moment, Tyndall (who loved controversy) became Mayer's new champion in Britain, going on the warpath against the established acceptance of Joule as the discoverer of the energy concept and the conservation law. Joule, supported by William Thomson and P. G. Tait, a professor of natural philosophy at Edinburgh, insisted that the proof required experimental work, while Mayer had been merely speculating. The heated controversy continued for some time, eventually dying down when the Royal Society awarded the Copley medal first to Joule in 1870 and then to Mayer in 1871.

The third scientist sometimes credited with the discovery of the conservation of energy was the versatile German physiologist and physicist Hermann Ludwig Ferdinand von Helmholtz (1821–1894). Born in Potsdam, to a father who taught him ancient languages as well as Italian, French, and Arabic and to a mother who descended from William Penn, the founder of the state of Pennsylvania, Helmholtz showed an early interest in physics. However, he went on to study medicine because the financial aid available in that field (in return for later military service) enabled him to afford a university education. After performing his required military duty as a surgeon and subsequently occupying professorial positions in physiology successively at the universities of Königsberg, Bonn, and Heidelberg, he was appointed to the chair of physics at the University of Berlin in 1871. Suffusing his approach with a broad and philosophical view, Helmholtz dominated German science during the second half of the nineteenth century.

Helmholtz conceived the principle of conservation of energy in 1847, two years after Joule had published his mechanical equivalent of heat. Unaware of Mayer's work, he had arrived at the same conclusion by similar physiological reasoning, but, using Joule's result, he managed to express the law more effectively and precisely than Mayer. Later called the first law of thermodynamics, it serves in part to define what is meant by energy and in part to prohibit certain processes: it forbids the construction of a *perpetuum mobile,* a machine that would be able to run indefinitely without a source of energy. Generations of clever but ignorant inventors have tried in vain to patent such miracle engines, and some occasionally do to this day.

Of course, the new science of heat was not confined to prohibitions. The French engineer Nicolas Leonard Sadi Carnot (1796–1832) was motivated by the desire to understand the basic principles underlying the steam engine and, if possible, to improve its efficiency. For this purpose he invented an idealized cyclical method for extracting work from the heat flowing from a substance at a higher temperature (a "reservoir") to another at a lower temperature. In his ideal cycle, a piston, driven by the expansion of a gas in a cylinder, performs work—like raising a weight—while extracting heat from the warmer and expelling some of it to the colder reservoir. Running the cycle backwards converts the heat engine into a refrigerator, which uses mechanical work to pump heat from the colder body to the hotter one. Close examination of this process shows that the efficiency of his engine—the ratio of the work it produces to the amount of heat it extracts from the hot reservoir—depends on nothing but the temperatures of the two reservoirs between which it operates, and that even under ideal conditions this efficiency is necessarily limited by the need to eject some of the heat extracted from the hot body into the cooler one as waste heat.

Sadi Carnot's invention of his idealized engine stimulated the discovery of what came to be called the second law of thermodynamics, which would eventually become the seed-bed for the conceptual revolution in physics leading to the demise of determinism. As Lord Kelvin formulated it, the second law says that for an engine such as Carnot's to work, the two substances between which heat is transferred must be at different temperatures. In other words, work cannot be obtained by extracting heat from a reservoir and letting some of it flow back into the same reservoir or into one at the same temperature. If this were not true, an engine could be built to run by extracting heat from the vast heat reservoir of the world's oceans and allowing some of it to leak back into the ocean. Such an engine would be called a *perpetuum mobile* of the second kind. Clausius, on the other hand, put the second law this way: There can be no process whose only effect is to transfer heat from a cold to a hot thermal reservoir. The two formulations were proved to be equivalent.

Lord Kelvin, named William Thomson at his birth in Belfast in 1824, was educated at home by his father, a professor of engineering at Belfast University. After his father accepted a professorship of mathematics at the University of Glasgow, young William entered that university at the age of ten to study natural philosophy, but he later changed to Cambridge University, where he graduated in 1845. After a year in Paris, he was appointed professor of natural philosophy at Glasgow, since Cambridge at that time had no such professorship; Isaac Newton notwithstanding, it was, with respect to science, still rather old-fashioned. In Glasgow he set up the first physics laboratory at any British university. In 1866 he was knighted and in 1892, while president of the Royal Society, he was given a peerage, taking the title Baron Kelvin of Largs. He died at Largs, Ayrshire, in 1907 and was buried in Westminster Abbey, next to Isaac Newton.

One of the most prominent physicists of the nineteenth century, Kelvin worked in many parts of physics, beginning with electricity and magnetism. He developed Faraday's discoveries in magnetism into a fuller theory, including oscillating circuits later used to broadcast radio waves, and prepared the ground for Maxwell and his pathbreaking theory of electromagnetism. Kelvin disliked Maxwell's theory, however, as he did not believe in the model its inventor had used as a scaffold. The first successful transatlantic telegraph cable was laid according to Kelvin's detailed specifications. But his most important contributions dealt with heat. Stimulated by the work of both James Joule, whom he had met in 1847, and Carnot—though he disagreed with the latter's acceptance of the caloric theory—he devised a new temperature scale on which the description of Carnot's cycle could be based.

At that time just as now, two different temperature scales were in general use. One had been devised by the Polish-born Dutch physicist Daniel Gabriel Fahrenheit (1686–1736), who set the zero point of his scale as the temperature of a mixture of sea salt, ice, and water and defined the normal human blood temperature to be 96° (now more accurately known to be 98.6°). The other scale we owe to the Swedish astronomer and physicist Anders Celsius (1701–1744), who chose the freezing point of water as 0° and its boiling point at sea level as 100° (32° and 212°, respectively, on the Fahrenheit scale; actually, Celsius designated the boiling point as 0° and the freezing point as 100°, but this was later reversed). This centigrade or Celsius scale is used in most of the world.

Both temperature scales have the peculiarity of entirely arbitrary zero points, which limits their scientific usefulness. For example, the gas law that Robert Boyle discovered stated that the product PV of the pressure P and the volume V of a gas is a constant if the temperature is held fixed. But as Charles and Gay-Lussac discovered, if the temperature is varied, that product increases linearly with the temperature. Kelvin, sticking to the size of each degree set by Celsius, proposed a scientifically more useful scale so that Boyle's law could be stated in the simple form $PV = RT$, which would be achieved by choosing as the zero point $-273.16°$ Celsius. In this form, the number R in Boyle's law is called the universal gas constant. While it implies that the volume of a gas shrinks to zero at 0° K, this implication is inconsequential, since no substance remains gaseous at very low temperatures. Kelvin's scale, indicated by the letter K, is now also called absolute temperature and is employed universally by physicists.

Kelvin played a less positive role in the Darwinian debate concerning evolution. Estimating the rate at which the earth had cooled since its initial formation, he came to the conclusion in 1862 that it could not be older than 400 million years, with 100 million as the most likely figure. Eight years earlier, Helmholtz had arrived at an estimate of about 25 million years, assuming that the heat of the sun was generated by gravitational contraction. Such erroneous estimates of the age of the earth severely constrained the time span available for evolution and, coming as they did from extremely reputable scientists, presented a formidable challenge to Darwin's theory. The reasons for Kelvin's and Helmholtz's gross underestimation of the earth's age were their ignorance of the nuclear reactions that were the real source of the sun's heat as well as their ignorance of the radioactivity that warms the interior of the earth. Both of these processes were discovered in the twentieth century.

The other originator of the second law of thermodynamics, Rudolf Julius Emmanuel Clausius, was born in 1822 in the Prussian city of Köslin (now Polish). He studied at the University of Berlin, obtained his doctorate at the University of Halle, and after teaching at the Royal Artillery and Engineering School in Berlin, he was appointed professor of physics at the newly established Polytechnicum in Zurich but returned to Germany twelve years later as a professor of physics at Würzburg and on to the University of Bonn, where he remained until his death in 1888. His last years were blighted by a painful wound he had suffered during the Franco-Prussian War—he had organized a volunteer ambulance service run by his students—and by the death of his first wife in childbirth, after which he took care of his six children by himself (remarrying two years before his death).

Clausius's most famous paper was published in 1850, shortly after he obtained his Ph.D. Rejecting the caloric theory and basing his work on that of Carnot and Lord Kelvin, he formulated the second law of thermodynamics in the form stated above, but in so doing he also introduced the concept of entropy, a name he coined from the Greek word for transformation. A change of entropy is defined as the heat flowing into a body divided by its temperature. In a Carnot cycle, which is reversible, the total change in entropy is zero, but when heat is simply allowed to flow from a hot substance to a cooler one, the total entropy increases—the hot substance loses less entropy than the cool one gains—whereas it would decrease if heat flowed from cold to hot. The second law of thermodynamics can therefore be stated in this form: The total entropy of an isolated physical system can never decrease; it can only increase or remain the same.

The rest of thermodynamics lays down detailed rules governing how the flow of heat through or into a body depends on the properties of the material making it up, the behavior of fluids when their temperature, pressure, or volume changes, and the way the state of a substance changes, from a gas to a liquid to a solid or vice versa, as its temperature or pressure changes. However, the formulation of the first and second laws is the most important achievement of thermodynamics. The first law differs from the individual laws of energy conservation that already existed within each of the other disciplines, such as mechanics, sound, and electromagnetism (though they had not really been recognized as such before the energy concept became clarified through thermodynamics) only by the wide reach of its sweep, its generality, and of course its inclusion of heat as a form of energy. The second law, however, introduced an entirely new notion into physics. In contrast to all other known physical principles, it contained an arrow of time, defined by the inexorable and irreversible increase in entropy. The concept of time's irreversible flow, popular and influential in philosophy and literature, had finally found its way into physics.

Both of these basic laws begged a fundamental question: if all matter is made up of particles subject to Newton's laws of motion, what is the explanation of the first and second laws and the other rules of thermodynamics? At this point they seem to be entirely ad hoc. What was needed was a precise, quantitative understanding of the nature of heat. Such an understanding, and, with it, an explication of the laws of thermodynamics, would lead to a complete upheaval of how physicists viewed nature. This is where probability entered into physics, replacing the strict determinism that had governed physical science up to that time.

But before pursuing that story, we have to turn to another important development in classical physics, a development that would both change and enlarge the scope of deterministic physics, carrying it all the way through the twentieth century, where it coexisted with the acausal quantum theory, each valid in its own sphere.

seven

Relativity

As enormously successful as Maxwell's theory of electromagnetism was, it contained an unexplained element: that all-pervading substance called ether, the vibrations of which manifested themselves as light. Maxwell's theory seemed to imply the reality of this mysterious stuff, by this point called the luminiferous ether, whose existence had been surmised, in one form or another, since Aristotle. The challenge of finding independent evidence for it was taken up by a team of two Americans, Albert Abraham Michelson (1852–1931, Polish-born) and Edward Williams Morley (1838–1923).

The reasoning underlying the Michelson-Morley experiment was very simple: if the ether filled the universe, presumably the earth rotated with respect to it and moved through it in its orbit around the sun. Since light moves at a fixed speed in the ether, just as sound does in air, the speed of a light beam as seen on the earth should vary depending on the direction of the beam in relation to the direction of motion of the laboratory on earth where it is measured. Because the speed of a point on earth relative to the ether, though unknown, was presumably extremely small as compared to the vast speed of light, it would of course be very difficult to detect the variation of the light speed with changing direction.

A master at employing interference effects of light beams as extremely sensitive sensors for the measurement of distances, Michelson, professor of physics at the Case School of Applied Science in Cleveland, Ohio, had originated a branch of experimentation that came to be called interferometry. The Michelson interferometer was its best-known device. In 1887 he and Morley set up a large platform on which two coherent monochromatic light beams, one of which moved along most of its way at right angles to the other, were brought together and made to interfere with each other. They hypothesized that the peaks of the waves of one would not be exactly at the position of the peaks of the other because of the difference in the time it took the two beams to traverse their paths, which in turn depended not only on the lengths of their paths but also on the different speeds during their journeys. The resulting peak shift would be visible as light and dark bands on a screen, just like Thomas Young's interference fringes. If the entire platform was then slowly rotated—it floated on a bed of mercury and, to prevent any shaking of the apparatus, the traffic in the streets around the laboratory was temporarily stopped—so that the two paths exchanged places, the resulting change in the observed interference pattern would enable the experimenters to infer the difference in speeds, and from this, the speed of the entire apparatus with respect to the ether.

The outcome of this elaborate experiment, variations of which have been repeated by others, was an extreme disappointment: as the apparatus rotated, the fringes did not budge. Though various implausible ad hoc explanations for this puzzling result were subsequently offered, the apparently nonsensical conclusion was unavoidable: no matter how fast or in what direction the laboratory in which it is measured moves, the speed of light is always the same. How could a light signal move with the same speed relative to two observers, one of whom moves along with it and the other moves in the opposite direction? It took a new physical theory to account for this strange conclusion: Einstein's theory of relativity. Pushing deterministic physics to its farthest limit, Einstein would at the same time replace Newton's monumental theory of gravity with a new structure.

Albert Einstein was born in 1879 in Ulm, Germany. Attending school in Munich, where the family had moved, and chafing under the discipline customary in German schools at the time, he showed no early promise. He received his higher education in physics and mathematics at the Eidgenössische Technische Hochschule (ETH) in Zurich, from which he graduated in 1900. After spending a year as a school teacher, he landed a job at the Swiss Patent Office in Bern and became a Swiss citizen. Soon afterward he entered a marriage that would end in divorce in 1919. Shortly after divorcing his first wife, Mileva, he married his cousin Elsa.

The year 1905 was Einstein's *anno mirabilis*—the analogue of Newton's plague years away from Trinity College—in which he published three profound papers. The first introduced the idea of "light quanta" (some twenty years later they were named photons) and thereby explained puzzling experimental results found by the German physicist Philipp Lenard (1862–1947) on the photo-emission of electrons by metals. The second explained Brownian motion as being the result of the irregular movements of water molecules. The third explained the negative result obtained by Michelson and Morley (though at the time he appears to have been unaware of that experiment, and his paper never mentioned it). The first paper would become a foundation stone in the revolutionary twentieth-century quantum theory; the second bolstered the molecular constitution of fluids; and the third would overturn our notion of space and time. In a fourth paper published the same year, he applied the results of the third to derive his famous equation $E = mc^2$, establishing the equivalence of mass and energy. Like Maxwell, Einstein thus straddled the great deterministic-probabilistic paradigm shift, making critical contributions to the completion of the old as well as to the birth of the new physics.

To the few physicists who read them, these papers initially appeared somewhat enigmatic. Nevertheless, they led to the offer of a junior professorship at the University of Zurich in 1909 and to a full professorship in Prague in 1911 and in Zurich in 1912. He moved to Berlin in 1914, where he was appointed director of the Institute of Physics at the Kaiser Wilhelm Institute and was freed of teaching duties so that he could pursue his research. In 1915 Einstein published his crowning achievement, the general theory of relativity. Even though the First World War had barely ended at the time, the British astronomer Arthur Stanley Eddington (1882–1944) led an expedition in 1919 to West Africa to observe to best advantage a solar eclipse from the island of Principe, in order to check out one of the predictions of general relativity: that as the light from a star passed near the sun, the force of the sun's gravity would bend it by a precisely calculated amount that could be observed during the eclipse. The verification of this prediction instantly made Einstein world famous, the successor of Isaac Newton as the architect of our understanding of the universe.

While the rest of the world celebrated the great scientist wherever he went on his travels, at home in Germany he was reviled as a Jew by the Nazis, whose ideology had attracted even some of his scientific colleagues. When Hitler came to power in 1933, Einstein happened to be visiting Western Europe and the United States, and he decided against returning to Berlin, where government thugs had already ransacked his home and his office while he was away. Resigning his directorship at the Kaiser Wilhelm Institute, he accepted a position at the Institute for Advanced Study in Princeton, New Jersey, which was established primarily to accommodate the famous man, and he remained there until his death in 1955.

According to Einstein's own recollection, the intellectual origin of the special theory of relativity went back to his youth, a time when he tried to imagine what it would be like to ride along on a wave of light. He decided that Maxwell's equations prohibited such a ride, since, no matter how fast he moved pursuing the light, he would always have to see it travel at the same speed. Such a conclusion required rethinking some basic and apparently obvious assumptions. If two observers in motion relative to one another see their clocks running at the same rates and their yardsticks to be of equal length, they cannot see the same light signal move at the same speed. Therefore one, or both, of these assumptions must be wrong; they had to be replaced by new ones: if you are moving with respect to me, I see your clock going slow and your yardsticks contracted, and symmetry demands that you see my clock going slow and my yardsticks contracted too. The reason why this had never been noticed before and is still difficult to notice is that these effects are minute unless our relative speed is close to the speed of light. There is no universal space in which the world moves along a universal time, as Newton had taken for granted. The standards of measurement for both space and time have to be separately established for each observer.

The special theory of relativity is based on the postulates that the speed of light has to be the same for all laboratories in uniform motion relative to one another, and that the laws of physics must have the same form in all of them—this had already been the case for Newton's laws of motion, but it is the universal nature of this "invariance" requirement for all laws, including those of Maxwell, that was new. Its essence is contained in a mathematical relation that expresses the location and time of every physical event as determined by an observer in the second laboratory in terms of those determined by one in the first. This relation is called the Lorentz transformation, because it had first been proposed by Hendrik Lorentz, who, however, had not drawn the correct physical conclusions from it.

Hereafter, and to this day, all physical theories dealing with velocities near that of light had to be Lorentz invariant: they must be such as not to change form when the location and time in them are subjected to a Lorentz transformation. (Einstein first wanted to attach the name "invariance" to his theory rather than "relativity," and he later regretted that it was the name "relativity" that stuck.) The theory also implied some important specific predictions, namely that no material body and no information can be transported faster than light and that mass and energy are equivalent, or convertible into one another, as expressed in the famous formula $E = mc^2$, in which c is the speed of light. This equation underlies the enormous energy released by a nuclear explosion, in which some of the mass in atomic nuclei is converted into energy.

Einstein's theory of 1905 came to be called the special theory of relativity because it was restricted to constant, uniform motion. However, it did not take Einstein long to begin thinking about more general, accelerated motion, realizing it would have to involve gravitation as well. Newton's force of gravity, after all, violated one of the prohibitions of his special theory: it was transmitted with infinite speed. Were the sun to disappear in a flash, the earth and the other planets would be instantly released from its pull and wander off on their own. So the special theory of relativity required that Newton's great universal gravitation theory be modified. This was an extremely demanding task, and it took Einstein ten years to fulfill it with the publication, in 1915, of the general theory of relativity. Its mathematical basis went back to the innovative work of several mathematicians of the nineteenth century: Gauss, Riemann, Bolyai, and Lobachevsky.

The special status of Euclid's fifth postulate (the parallel postulate), which implied that the interior angles of a triangle add up to 180°, had been recognized for over two millennia, but no mathematician had been able to prove it from the other four postulates, or to do geometry without it. And yet it always felt extraneous. Why was nature constructed according to Euclidean geometry? The eighteenth-century German philosopher Immanuel Kant, believing it to be a necessity for human thought, saw it indelibly imprinted on our description of nature. When, in the nineteenth century, some courageous mathematicians began to consider whether geometry could do without the infamous fifth postulate, the mathematical community looked askance. The great Gauss, who had made profound studies of the geometrical properties of curved surfaces, was intellectually well situated to judge this question, and he came to the private conclusion that a geometrical system without the parallel postulate would indeed be possible. Conservative and reticent as he was, however, he filed his work away and did not utter or publish a word.

Born in 1802 in Kolosszvár, Hungary, János Bolyai was the son of Farkas Bolyai, a mathematician and friend of Gauss's, who was obsessed with the problem of proving Euclid's fifth postulate from the other four. Even after he thought he had discovered such a proof and sent it to Gauss, who had found a flaw in his argument, he continued his futile quest. Meanwhile, young János had grown up with a great talent for the violin and mathematics, but had entered the army engineering corps. Nevertheless continuing to pursue his mathematical interests, he came to the firm conclusion that his father was chasing a mirage: the proof he sought would remain impossible. By 1823 he had succeeded in constructing a geometry that violated Euclid's parallel postulate: in the plane of a given point P and a given line L, a whole fan of lines through P would be possible without ever intersecting L. The opening angle of this fan characterized the geometry as a parameter; if this parameter was allowed to shrink to zero, his geometry would become Euclidean. He had found the first non-Euclidean geometry, and he wrote it up in a paper entitled "The Absolute True Science of Space."

When János's skeptical father sent the paper to Gauss, the reply they received after a considerable delay astonished them both: Gauss had been thinking along similar lines for over thirty years and had come to the same conclusion as János. The young Bolyai felt deprived of the pride of priority, but since Gauss had published nothing about it, he saw no obstacle to his own publication. János's paper was finally printed in 1832 in the form of an appendix to an article by his father. It remained the only work János Bolyai ever published in his lifetime, and not even under his own name. To the deep disappointment of both father and son, the paper received little attention, and they retreated dejectedly from the world, living in the same house for a while. János eventually fathered three children and died in 1860, four years after his father.

The second, independent discoverer of non-Euclidean geometry was the Russian mathematician Nikolai Ivanovich Lobachevsky, born in 1792 in Nizhni Novgorod (now Gorki). His father, a low-level government clerk, died when Nikolai was a child, leaving his mother to raise him alone and provide his early schooling. At the age of fourteen Nikolai entered the University of Kazan, where he studied mathematics and remained as a librarian, professor, dean of physics and mathematics, and eventually as rector until 1846. In 1832 he married a wealthy woman with whom he had seven children, and Tsar Nikolas I raised him to the hereditary nobility in 1837. During his last years he was nearly blind from cataracts in both eyes, and he died in 1856 in Kazan.

Lobachevsky's and Bolyai's discoveries of the same kind of geometry overlapped in time. After fruitlessly attempting, like Bolyai, to prove the fifth postulate from the others and abandoning that search, Lobachevsky presented an outline of his new geometry in 1826 to his colleagues in Kazan and published his first two papers on it in 1829 and 1830 in the *Kazan Messenger*. In his geometry, as in Bolyai's, the interior angles of triangles add up to a total of less than 180°. (This kind of geometry is now called hyperbolic; Lobachevsky called it "imaginary.") Its existence demonstrates that Euclid's fifth postulate was independent of the others and could be altered with no harm. Like Bolyai, Lobachevsky never received any recognition for his pathbreaking work during his lifetime. And neither Bolyai nor Lobachevky proved that their geometry might not contain internal contradictions. That proof of consistency was supplied in 1860 by the Italian mathematician Eugenio Beltrami (1835–1899).

A different kind of non-Euclidean geometry was discovered by Bernhard Riemann, a mathematician of much wider sweep and influence than either Bolyai or Lobachevsky. The shy and frail son of a Lutheran pastor, Georg Friedrich Bernhard Riemann was born in 1826 in Breselenz, Hanover, Germany, and showed a distinct mathematical aptitude early on. At the age of nineteen he entered the University of Göttingen to study theology but soon switched to mathematics. Since the sole notable mathematician in Göttingen at the time was Gauss, who taught only elementary classes, he went to the University of Berlin after one year. But then he returned to Göttingen to study under the aging Gauss, writing a brilliant doctoral dissertation on what came to be called Riemann surfaces. To qualify for a university career, he submitted a *Habilitationsschrift* on Fourier series but, at the suggestion of Gauss, he also studied the foundations of geometry. The rest of his short life he spent as a professor at the University of Göttingen, eventually taking the place of Gauss after the latter's death. At the age of 36 he married Elise Koch and fathered a daughter. That same year, however, he came down with pleuritis and subsequently developed tuberculosis, the same disease from which his mother and four of his siblings had died. To recuperate, he repeatedly traveled to Italy, both for the art he greatly admired and the climate. He died in 1866 at Selasca on Lago Maggiore.

The influence of Riemann on mathematics and mathematical physics is impossible to overstate. Rather than a developer of formalisms, he was a profoundly conceptual thinker. The entire large subject known as the theory of analytic functions is based upon his ideas and Cauchy's. The Riemann hypothesis, connecting the zeros of the Riemann zeta-function, which is defined in terms of an infinite sum with no apparent connection to the theory of numbers, with the distribution of prime numbers among the integers, is to this day perhaps the deepest unsolved puzzle in mathematics. As the mathematician David Hilbert put it, "If I were to awaken after having slept for a thousand years, my first question would be: has the Riemann Hypothesis been proven?"

But Riemann made his most profound contributions to the field of geometry, especially to what is now called differential geometry, a subject he approached philosophically and not just technically, taking into account the possible interaction between physical objects and the space in which they are situated. This idea would have an important influence, half a century later, on the thinking of Albert Einstein. What is more, in the paper on the foundations of geometry that he wrote at Gauss's suggestion as part of his *Habilitation*—unpublished until a year after his death—he discovered a non-Euclidean geometry that differed from the Bolyai-Lobachevsky version. In this geometry, every straight line in the plane of a given line L intersects L, and the sum of the angles in a triangle is greater than 180° (how much greater depends on the size of the triangle). This kind of geometry is now called elliptic or Riemannian. Both elliptic and hyperbolic geometries are characteristics of curved spaces (the concept of curvature had been precisely defined by Gauss); Euclid's fifth postulate is valid only if the space is flat, that is, its curvature is everywhere zero.

These ideas are all most easily visualizable for two-dimensional spaces (surfaces). The surface of a sphere is a Riemannian space of constant positive curvature, and great circles are straight lines. (A ship sailing a straight course is really following a great circle on the ocean.) Any two of these straight lines necessarily intersect somewhere, and the interior angles of a triangle made up of pieces of great circles add up to more than 180°. For example, you can form a large triangle on the surface of the earth whose sides consist of part of the equator and the two meridians of 0° and 90° longitude; all three of its interior angles are 90°, adding up to 270°. A much smaller triangle drawn in the sand, whose sides of course are also pieces of great circles, has interior angles that add up to only 180°. Thus in a curved space, the sum of the angles in a triangle depends on its size: the geometry in a region small enough compared to the curvature of the space is always approximately Euclidean. The geometry on a surface of constant negative curvature, on the other hand, is hyperbolic.

Such a surface looks like two welded-together straight bugles with infinitely long stems and is called a pseudosphere.

With this background in mind, let us return now to Einstein and his struggle to construct a form-invariant theory of gravitation, that is, a theory in which the equations take the same form in any physical laboratory, even in one that is moving or accelerated by the force of gravity. In such a theory the action of gravitation would not be instantaneous, as in Newton's. During the year 1912, after he had moved to Zurich, Einstein wrestled most intensely with the problem, hampered by his limited knowledge of the needed mathematical tools. He finally saw the light with the help of a mathematician, his best friend, Marcel Grossmann, who had been instrumental in bringing him back to Zurich. As Einstein described in a talk he gave ten years later in Kyoto, "If all [accelerated] systems are equivalent, then Euclidean geometry cannot hold in all of them . . . This problem remained insoluble to me until 1912, when I suddenly realized that Gauss's theory of surfaces holds the key for unlocking this mystery . . . However, I did not know at that time that Riemann had studied the foundations of geometry in an even more profound way . . . I realized that the foundations of geometry have physical significance . . . So I asked my friend [Marcel Grossmann] whether my problem could be solved by Riemann's theory."[1] The way now lay open for him to express his theory by means of the tensor calculus, a branch of mathematics that few physicists at the time were familiar with.

By 1915 he had it all wrapped up and the general theory of relativity was published, though almost no one understood it. Nevertheless, it led to two specific astronomical implications: the bending of light by gravity, which Eddington verified on his expedition to Africa in 1919, and an explanation of the precession of the axes of the planet Mercury's orbit, an unsolved puzzle dating from the beginning of the nineteenth century. In the general theory of relativity, Newton's gravitational field is replaced by the influence that any object exerts on the geometry of all its surroundings: the mass of a large star bends the very space of the universe. And, rather than being pulled by the force of gravity, other objects respond by simply moving in this curved space along geodesics, that is, along lines that are "straight" in the resulting Riemannian geometry.

This theory resolves what had always been a bit of a mystery in Newton's equations: the equality between the inertial mass of an object on one hand, such as the mass that resists a change in velocity, which appears on the right-hand side of the equation $F = ma$, and, on the other hand, its gravitational mass, to which the force of gravity is proportional. Simply an unexplained ad hoc assumption in Newton's physics, this equality has the curious result that the motion of any object subject to nothing but gravity is completely independent of its mass. As this fact is extremely well verified by elaborate experiments, Einstein made it a cornerstone of the general theory of relativity; without this principle of equivalence, his geometrization of gravity would have been impossible.

Difficult as Einstein's equations are to solve in general, the German astronomer Karl Schwarzschild (1873–1916) found an exact spherically symmetric solution of these equations, thereby greatly facilitating the study of some of the theory's consequences. One of these turned out to be a singularity into which, as Robert Oppenheimer pointed out, every sufficiently massive object would irretrievably collapse. John Wheeler called them black holes, as not even light could escape from them, and the name stuck. Nothing approaching a black hole to within its Schwarzschild radius could get away from its gravitational pull. (This statement would later be somewhat modified by the English physicist Stephen Hawking, b. 1942, on the basis of quantum mechanics.) Although general relativity is essential for the details of this phenomenon, already in the late eighteenth century Laplace had remarked that the gravitational pull of a sufficiently small and massive stellar object would have an escape velocity—the speed required to overcome its gravity so as to get away from it—greater than the speed of light. As it therefore could not emit light, it would be invisible but might be detected by having planets or other stars circling about it. On the basis of just such evidence, astronomers now believe they have found these objects, often at the centers of galaxies, including our own galaxy, the Milky Way.

Another consequence of replacing Newton's action-at-a-distance gravity by a field theory is that, just as Maxwell's theory implied electromagnetic waves, Einstein's implied gravitational waves. These would be very difficult to detect, and, in spite of a number of attempts and some erroneous claims of success, direct evidence for them has not yet been found. However, in 1975 the American physicists Russell Alan Hulse (b. 1950) and Joseph Hooton Taylor, Jr. (b. 1941) discovered indirect evidence in the behavior of two pulsars orbiting one another. Details of the orbits of these two neutron stars could be well explained by their emission of gravitational radiation.

Einstein's gravitational theory represented the culmination of all of deterministic physics, and the problem that the twenty-first century inherited was to make this theory come to terms with the probabilistic revolution in physics that pervaded the nineteenth and twentieth centuries. In the meantime, however, general relativity served as the backbone of the entire structure of twentieth-century cosmology, which began with the American astronomer Harlow Shapley (1885–1972), who introduced a new understanding of galaxies and their distances. Following detailed measurements on the relatively rare variable stars among globular clusters (roughly spherical systems of tens of thousands of stars, gravitationally bound together) he had performed at the newly constructed 100-inch reflecting telescope at Mount Wilson Observatory near Pasadena, California, he devised a method of determining distances across vast interstellar spaces. This he did by using a connection between the luminosities of the "Cepheid variables" and the periods of their varying light outputs. (Luminosity is the intrinsic brightness of a star, as distinct from its apparent brightness as seen by us at a distance. Knowing both the apparent brightness of a Cepheid variable and, on the basis of its period, its intrinsic luminosity allowed Shapley to use the inverse-square law to calculate how far away it was.)

Henrietta Leavitt (1868–1921), a tireless "computer" working at the Harvard Observatory, had discovered that the periods of Cepheid variables were correlated with their apparent brightness. The final step was taken by the Danish astronomer Ejnar Hertzsprung (1873–1967), who used the motion of the entire solar system, first observed by William Herschel, as a basis for a parallax calculation that established the connection between Cepheids' period and their intrinsic luminosities. They could now serve as cosmic milestones to measure the distance to other stars and galaxies.

Astronomers employ three different units to state such distances. The first is the astronomical unit, or AU, which is approximately the average distance between the sun and the earth in its elliptical orbit (about 93 million miles); the second is the parsec, abbreviated as pc and defined as how far from us an object would be if it produced a parallax of 1″ (one second of arc) when seen from the sun and the earth; the third is the light year, the distance that light travels in one year (1 pc = 3.262 light years). The custom of stating astronomical distances in terms of light years—at least when talking to the public—seems to have been introduced by Edwin Hubble in the 1930s.

This new means of determining interstellar distances enabled Shapley to radically increase astronomers' estimate of the size of our own galaxy, the Milky Way, giving it a diameter of some 300,000 light years. Furthermore, he calculated the distances to some spiral nebulae well outside our own galaxy. (At that time, the word *nebula* was applied to any stellar object that was not simply a star.) That these spiral nebulae were actually huge star systems—galaxies like the Milky Way—was shown in 1924 by Hubble.

Edwin Powell Hubble was born in 1889 in Marshfield, Missouri. When he was eight, the family moved to a suburb of Chicago, first Evanston then Wheaton, where he graduated from high school at sixteen, excelling in both sports and academics. He went on to the

University of Chicago, earning a B.S. in mathematics and astronomy and also doing well as an amateur heavyweight boxer. Awarded a Rhodes Scholarship, his interests changed to law and he obtained a degree in jurisprudence at Queen's College, Oxford, in 1912, at the same time continuing his endeavors in athletics, particularly in track and field events. After returning to the United States, he practiced law in Louisville, Kentucky, for a year but, becoming bored with it, he re-entered the University of Chicago to earn a Ph.D. in astronomy, doing research for the degree at the university's Yerkes Observatory at Williams Bay, Wisconsin.

During the ensuing war he served in the infantry in France, and in 1919 he obtained a post as astronomer at the Mount Wilson Observatory, where he remained an active observer of galaxies for the rest of his life. During World War II he served as a ballistics expert for the U.S. War Department. Hubble died in 1953 in San Marino, California. NASA's first telescope placed in orbit around the earth in 1990 was named after him.

After discovering galaxies external to our own and proceeding to classify them according to their shapes and velocities, Hubble made the remarkable discovery in 1927 that all these external galaxies seemed to be receding from the Milky Way. The method for determining the speed with which a star is moving away from us, or toward us, is based on observing the spectral composition of its light. Every element, when heated, emits light composed of a characteristic set of colors, which, when sent through a prism, separate themselves into spectral lines. Therefore, the elements making up any given star can be identified from its spectrum. What Hubble found was that all the spectral lines from stars in far-away galaxies were shifted toward the red end of the spectrum, meaning that their wavelengths were stretched. The obvious interpretation of this red shift was that it represented a Doppler shift analogous to the lowering of the tone of a receding siren. Conversely, if these galaxies had been moving toward us, their spectra would have been blue-shifted, just as the sound of a siren approaching us seems to have a higher pitch.

Hubble's interpretation of the spectral shifts of galaxies was not immediately accepted. Some astronomers believed the red shift was caused by something in the interstellar medium traversed by the light. As late as 1936 Hubble himself harbored doubts about the Doppler shift explanation for red shift—and hence also about the recession of the galaxies.[2] What is more, Hubble not only discovered that all the galaxies are moving away from us, but in 1929 he found that the speed at which a galaxy was receding was proportional to its distance from us. The constant of proportionality is now called Hubble's constant, and it is one of the basic parameters of the universe (despite the fact that Hubble's own calculation of the numerical value of this constant turned out to be incorrect).

The idea that the universe should be constantly expanding not only contradicted the earlier universal view that the cosmos was static, but it had a number of remarkable consequences. Since the velocity of recession of a galaxy is proportional to how far away it is, at some great distance away the speed of recession starts to approach and even equal the speed of light. Therefore, the light emitted by stars farther away from us than this Hubble radius can never reach us. The Hubble radius, then, defines the radius of the knowable universe.

Hubble's discovery of the expansion of the universe had an immediate effect on the theory employed to explain the structure of the cosmos, the general theory of relativity. When Einstein, shortly after developing the theory, began to apply it to cosmology, he found that his equations had no solutions that described a static, closed universe of fixed dimensions, with relatively slowly moving stars in it, as the cosmos was thought to be at that time. To adjust his theory to accord with this assumed fact, in 1917 he introduced an artificial fudge factor in his equations, which came to be known as the cosmological constant.[3] However, when Hubble discovered that the universe is expanding in size, Einstein removed the ugly fudge factor, calling it "the greatest mistake in my life." Ironically, the cosmological constant was resurrected some 75 years later for other reasons, as we will see.

Another consequence of an expanding universe concerned its beginning. The Belgian Jesuit priest and cosmologist Georges Edouard Lemaître (1894–1966), who had become fascinated with Einstein's theory and studied it deeply, had arrived at the independent conclusion that the general theory of relativity implied that the universe must be expanding, as Hubble confirmed. But how and when did this expansion start? After becoming professor of astrophysics at the University of Louvain, he published his *Discussion on the Evolution of the Universe* in 1933, which contained his theory that the universe began between 20 and 60 billion years ago with what was subsequently called a Big Bang, which he visualized as an explosion of a "primal atom." While the primal atom has been discarded, the Big Bang, improved in 1946 by the Russian-born American physicist George Gamow (1904–1968), is still the generally accepted explanation for the beginning of the cosmos among physicists.

Lemaître's was not the only model of the universe constructed by means of the general theory of relativity. Another model, employing a different geometry, was built in 1917 by the Dutch mathematician and astronomer Willem de Sitter (1872–1934), and a third by Arthur Eddington in 1930. These all differed in the overall curvature they assigned to space in their universes, and which of them is closer to reality has not yet been definitely decided by observation.

The twentieth century generated new knowledge and theories of many other astrophysical aspects of cosmology, but those theories rely on quantum mechanics, the culmination of the revolutionary change in physics from determinism to probability, and their description will have to wait until Chapter 11. We will first turn to the beginning of this paradigm shift, statistical mechanics.

eight

Statistical Physics

That the heat content of a body consisted of nothing but the motion of its constituents was not a new notion at the beginning of the nineteenth century. Many versions of it had been around at least since the time of Robert Boyle, who regarded the springlike pushing by the molecules of a gas as the cause of the pressure the gas exerted on the walls of its container. These theories tended to suffer under the influence and authority of Isaac Newton, who declared at one time that the molecules of a gas attract one another, at other times that they repel one another, and at still others that they are simply independent hard spheres.

The initiator of the modern kinetic theory was John Herapath (1790–1868). A mathematical prodigy, he grew up in Bristol with little formal schooling but, after learning French at an early age, educated himself by studying the works of the great French mathematicians. He then decided to become a teacher of mathematics, setting up his own school. Although he published a two-volume treatise entitled *Mathematical Physics,* he always remained a scientific amateur and eventually turned to a successful career in journalism as the editor of *Herapath's Railway Journal.*

Herapath's scientific interest began with Newtonian celestial mechanics and from there shifted to heat and gases in a futile attempt to explain gravity without recourse to action at a distance. Considering the absolute temperature of a gas as proportional to the momentum of its constituent molecules, that is, proportional to the product mv of their mass m and velocity v, he managed to derive a version of Boyle's gas law that had been found by Bernoulli in 1738 (though Herapath did not know of Bernoulli's work), $PV = \frac{1}{3}Nmv^2$ where N is the number of molecules in the volume V. This, of course, meant that Herapath's version of the gas law had PV proportional to T^2, contradicting the Gay-Lussac version, according to which PV was proportional to T. The paper containing these results was rejected in 1820 by the Royal Society but eventually published in the *Annals of Philosophy.* Herapath's book *Mathematical Physics* contained more details of the derivation as well, and it was read with interest by Joule when he developed the mechanical equivalent of heat. His work was thus noticed by some of the important creators of thermodynamics, but it was otherwise forgotten.

The next step in the kinetic theory was taken by another amateur scientist, John James Waterston (1811–1883), who was born in Edinburgh. Waterston studied civil engineering, but at the University of Edinburgh he also attended lectures in mathematics and physics as well as chemistry and anatomy. Eventually, he became a naval instructor with the East India Company in Bombay and held this position until he resigned in 1857, returning to Scotland, where he devoted his life to scientific work and lived on his savings.

Waterston's interest in the behavior of gases, like that of Herapath, evolved from trying to explain the action of gravity by means of the push of particles. In a book entitled *Thoughts on the Mental Functions,* published in 1843, he outlined his ideas on kinetic theory, apparently unaware of either Herapath or Bernoulli. These ideas included a connection between the temperature of a gas and the vis viva, that is, mv^2 of its molecules. The correct expression for the kinetic energy, or vis viva, is ½mv^2, but the needed factor of ½ was not always included. A long paper he sent to the Royal Society in 1845 was more explicit: on the basis of a model of a gas consisting of elastic spheres that collided with one another after traveling freely for a certain distance, the absolute temperature should be proportional to mv^2.

As a consequence, he fashioned a law whose generalization later came to be known as the equipartition theorem: in a mixture of gases consisting of molecules of different weights, the mean-square velocity (that is, the average of v^2) of the molecules of each of them should vary inversely with their densities. This is so because all of the gases in the mixture should have the same temperature, and hence the same average mv^2. (Notice here the first mention of averages.) The Royal Society not only rejected it (the two referees regarded it as nonsense) but refused to return the paper, of which he had retained no copy. After Waterston's death, Lord Rayleigh found it in the archives of the Royal Society. His work remained unknown during his lifetime, his influence nil. The book, with its uninformative title, remained unread until J. B. S. Haldane discovered it and had it reprinted in 1928.

On the other hand, the German chemist August Karl Krönig (1822–1879), a high school teacher who edited *Fortschritte der Physik,* seems to have received more attention for a short paper he published in 1856 in *Annalen der Physik* than is warranted by the originality of its content; though independent of the work of Herapath and Joule, it made no real advance over their work. Rudolf Clausius was the one who carried the ball farther, making the kinetic theory a substantial part of thermodynamics, to which he had contributed in an important way. His paper, "Ueber die Art der Bewegung, welche wir Wärme nennen" ("About the kind of motion we call heat"), published in 1857, made the final, decisive break with the caloric theory, presenting the molecules of a gas as moving freely, with no forces acting between them (some other physicists had seen the molecules as vibrating about fixed centers), the temperature being a measure of the vis viva of their motion.

What is more, the molecular motion was made up of more than straight-line movements but included the possibility of rotations and, for gas molecules containing more than one atom, internal vibrations of these atoms relative to one another. Recognizing radiant heat as similar to light (this was before Maxwell's theory of light), he envisioned a simple transfer of the vibratory motion of the ether to the kinetic energy of moving molecules as the explanation of how the absorption of heat radiation managed to warm up a gas. For Clausius, this was the final proof that the caloric theory was inadequate.

The "ideal gas" described by Clausius consisted of molecules which, though not just points, occupied an extremely small part of the total volume of the gas and were within the ranges of one another's forces during collisions for a very short time only; otherwise they exerted no influence on one another. In a liquid, on the other hand, the molecules were never far enough apart to escape the influence of their mutual forces, and in a solid they vibrated or rotated about fixed equilibrium positions. In his calculations Clausius generally attributed to each molecule the average velocity they had in the gas, paying no attention to deviations from this average. However, he based his explanation of the phenomenon of evaporation of a liquid on the assumption that, whereas the average velocity of the molecules was too low for them to overcome the attractive forces between them, a few deviated far from the average, moving fast enough to be able to escape from the surface. Only beyond the boiling point would they all move so fast that the intermolecular forces could no longer hold them together and could from then on be neglected. To accomplish this change of state from liquid to gas (and similarly, from solid to liquid) required the addition of a certain amount of extra heat, called latent heat, the extra energy needed to overcome the mutual attractions.

One of the objections raised against the final version of the kinetic theory, in which molecules moved freely without any forces between them except at very short distances, was that it seemed to imply a very rapid diffusion of a new gas entering into a mixture. The speed with which molecules moved could be estimated using the Bernoulli version of the gas law, and it came out to be on the order of about 2,000 feet per second. But in fact, a new odor introduced into a room was not noticeable that quickly at a distance. Clausius's answer was to introduce the concept of the mean free path, the average distance that a molecule travels in a gas before being deflected from its original direction by hitting another. As a result of such collisions, the actual zigzag path of a molecule from one end of a room to the other is very much longer than a straight line, and diffusion is considerably slower.

Given the size of a molecule, its mean free path in a gas could be calculated or it could be inferred from other measurements and then the size of molecules could be deduced from it. This is how the Austrian chemist and physicist Joseph Loschmidt (1821–1895) estimated the size of molecules in 1865. Deducing the mean free path from measurements of viscosity and calculating the fraction of the volume of a gas occupied by the molecules themselves from a comparison of the densities of the same substance in gaseous and liquid form, he concluded that the diameter of a molecule of air is about 10^{-7} cm (this is about four times too large, but a respectable estimate) and that the number of molecules of an ideal gas in a cubic centimeter is $N = 2 \times 10^{18}$, which is sometimes called Loschmidt's number. (Its presently accepted value is 2.687×10^{19}.)

While the contributions of Clausius to the kinetic theory of gases were important, they stopped short—notwithstanding his mean-free-path argument—of the steps taken by Maxwell and Boltzmann, who initiated the revolution brought about by the introduction of the concept of probability into physics. Clausius almost always assumed that all gas molecules moved with their average speeds. "In reading Clausius we seem to be reading mechanics," wrote J. Willard Gibbs; "in reading Maxwell, and in much of Boltzmann's most valuable work, we seem rather to be reading in the theory of probabilities."[1] In view of the size of the numbers of Avogadro and Loschmidt, it was, of course, inevitable that probabilistic considerations would enter when the behavior of macroscopic bodies had to be understood in terms of their microscopic constituents. To follow the individual paths of a billion trillion molecules was quite impossible and not really relevant for the task of explaining how a gas behaved. The focus of Herapath and Waterston had still been on shooting down the caloric theory and establishing the kinetic theory in its place. For Maxwell, that issue was settled, and he endeavored to understand the consequences of the kinetic theory in all its details. Probability was at that point unavoidable, and there was nothing inherently revolutionary about using this concept in physics, nor did Maxwell feel like a revolutionary while the story slowly unfolded. The real extent of the resulting paradigm shift would not become apparent until after the work of Boltzmann.

In contrast to Clausius, Maxwell was not satisfied with imprecise, qualitative statements like some molecules of a gas at a given temperature moved faster, and some more slowly, than the average. He wanted to know exactly how their various velocities were distributed around the average. On the basis of a few simple, reasonable assumptions (that all directions of motion should be on the same footing and that positive and negative deviations from the average were equally likely), in 1860 he derived—and in 1867 he improved—a formula of a similar form as the one Laplace had found for a normal distribution of errors. If the gas has the temperature T, the probability for the speed of a molecule to be found between v and $v + d$, for small increments d, can be plotted, as shown in Fig. 12. It has its maximum at $v = \sqrt{2kT/m}$, where m is the mass of the molecules and k is a fundamental constant, later called the Boltzmann constant. In other words, the most probable speed of a molecule in the gas increases as the square root of the temperature. Boltzmann subsequently gave a more general argument for this Maxwellian distribution law, the first appearance of an explicit probability distribution in physics.

As one of the applications of his distribution law of molecular speeds, Maxwell calculated the viscosity of gases at various temperatures and pressures, and he and his wife then went ahead and performed the actual measurements in 1865. In agreement with his prediction, they found that over the ranges they had studied, the viscosity of a gas is independent of its pressure. Many other applications followed, including a derivation of Dalton's law of partial pressures and investigations of heat conduction. Among Maxwell's last innovations before his death in 1879 was the introduction of the notion of ensemble averaging as a device for solving statistical problems.

From a probabilistic point of view, the problem was to understand

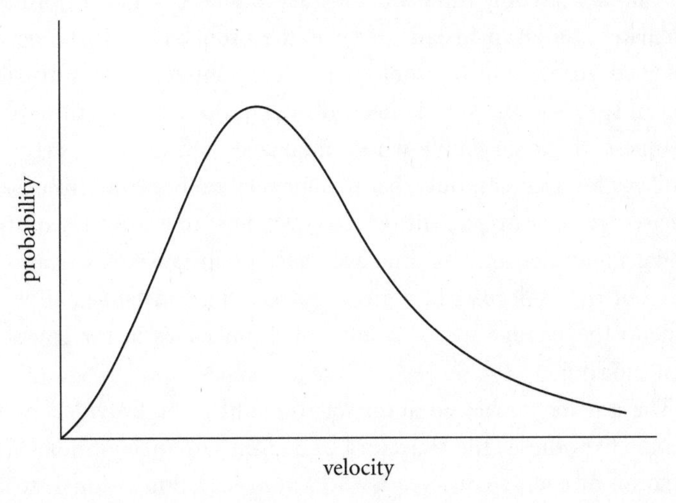

Figure 12 A Maxwellian velocity distribution.

the meaning of the averages and of the probabilities that played such an important role in going from microscopic to macroscopic physics. Maxwell's technique was, first of all, to employ Hamilton's formulation of the Newtonian equations, which used both the position coordinates and the momenta of particles as the fundamental variables to be tracked during their motion. This had the advantage that knowing both positions and momenta of particles at one instant allowed the complete prediction of their future motion. The precise specification of the state of a system consisting of n particles therefore required $6n$ numbers (three each for the coordinates of their locations and of their momenta): their motion had to be described as taking place in a space of $6n$ dimensions, called the phase space. The motion of all the n molecules of a gas was thus described as a simple trajectory in a $6n$-dimensional phase space, and at most one such trajectory passed through any given point in that space.

The next step was to introduce an ensemble, an infinite collection fictitiously made up of systems all of the same kind and the same energy but with different initial conditions (different starting positions and momenta compatible with the same total energy). In order to calculate, at a given instant of time, the probability of finding any specified property of the molecules—for example, a given distribution of speeds—one simply needed to determine the fraction of the members of the ensemble whose molecules had that property, and similarly for averages. But what was the relation between this ensemble average (a term later introduced by Gibbs) of a property of individual molecules and its time average? For instance, is the average speed of the molecules of a given gas at a fixed instant of time the same as the average speed of individual molecules in the course of their motion?

The answer to this question was thought to be provided by the ergodic hypothesis: the trajectory of a given system, confined by the equation of energy conservation to a $(6n - 1)$-dimensional surface in phase space, would eventually pass through every given point on that surface. This assumption quickly turned out to be impossible and had to be replaced by the quasi-ergodic hypothesis, which would be sufficient: given any area, no matter how small, on the energy surface, every trajectory on this surface eventually passes through it. Another way of saying the same thing is that every trajectory on a given energy surface eventually covers this surface densely. Obviously, not every mechanical system is ergodic in this sense. A planet orbiting the sun remains confined to a one-dimensional orbit. However, Poincaré proved that almost all mechanical systems of sufficient complexity are, in fact, quasi-ergodic. (Mathematically speaking, "almost all" means that the exceptions have probability zero. Since the set of mechanical systems is infinitely large, probability zero does not mean there are no exceptions. The required complexity is not very large; systems consisting of three particles are complex enough.)

Ergodicity alone, it was eventually realized, was not sufficient to assure the equality of ensemble averages and time averages, but the additional assumptions needed were later supplied. The first to do this, in 1911, was the Austrian-born Dutch physicist Paul Ehrenfest (1880–1933), assisted by his wife Tatyana, a mathematician. A simpler procedure, and one often adopted nowadays, is to assume for statistical purposes that the probability of finding an individual system in a given region of the energy surface is proportional to the area of that region, and that all systems within that region have equal a priori probabilities. It should be clear by now that Maxwell's introduction of probabilities had opened a can of worms, but there was no way of getting them back into the can.

The second physicist responsible for the probabilistic revolution in physics was Josiah Willard Gibbs, the first American theoretical physicist to acquire an international reputation. Born in 1839 in New Haven, Connecticut, and educated at Yale, where his father was a professor of sacred literature, Gibbs received his Ph.D. in 1863. Three years later, after both of his parents and two of his sisters had died, he traveled to Europe, accompanied by his two remaining sisters, and spent a year each at the universities of Paris, Berlin, and Heidelberg, attending lectures and reading widely in mathematics and physics. Upon his return to America, he settled in New Haven in the house where he had been born, never again to leave the United States and rarely even New Haven. He never married, sharing the house with his two sisters and the family of one of them. Appointed professor of mathematical physics at Yale (though for nine years he did not receive a salary and supported himself with his inheritance), he remained there until his death in 1903.

Though Maxwell had been the first to introduce probabilities and ensembles, Gibbs was the real founder of the science of statistical mechanics. He published his first two papers on thermodynamics in 1873 and followed them shortly with a major work on the subject, giving the concept of entropy much more prominence and power, as well as clarity, than anyone before him, including Clausius himself, who had introduced it. The venue for these publications was the *Transactions of the Connecticut Academy of Arts and Sciences,* a provincial journal of small circulation, but Gibbs sent individual copies to a large number of prominent European scientists. One of his readers turned out to be Maxwell, who thereupon became one of Gibbs's strongest and most enthusiastic supporters in Europe. Nevertheless, he remained relatively unknown on the continent until the chemist Friedrich Wilhelm Ostwald, who regarded Gibbs's results on chemical equilibrium as fundamentally important for his field, translated his papers into German, and Henri Le Châtelier, who thought they were comparable to Lavoisier's work, translated them into French.

A year before his death, Gibbs published his book *Elementary Principles in Statistical Mechanics Developed with Special Reference to the Rational Foundations of Thermodynamics* (the term *statistical mechanics* was his coinage), which summarized his contributions. Deliberately leaving the nature of the constituents of gases open, he declared, "Certainly, one is building on an insecure foundation, who rests his work on hypotheses concerning the constitution of matter."[2]

Instead, he schematically dealt with the statistical behavior of systems whose number of "degrees of freedom"—essentially the number of coordinates required to specify them—was of the same large order of magnitude as the number of molecules in a gas.

Gibbs was fully aware of the gaps left in the "rational foundations" that he was unable to fill. Even though he treated the thermodynamic approach to equilibrium from the statistical point of view, he did not really explain the mystery of irreversibility, that is, how the second law of thermodynamics could contain an arrow of time—the always increasing entropy—when the underlying equations of motion of the constituent particles were completely reversible, with no such directionality. It was Boltzmann who explicitly provided the solution to this puzzle, though after a struggle that exacted a heavy personal price. Even though Gibbs credited Boltzmann in his book as one of the founders of the statistical approach, he did not mention his solution to the arrow-of-time puzzle.

Ludwig Eduard Boltzmann, born in Vienna in 1844, received his elementary education from a private tutor at home and his secondary schooling in Linz after his family moved there. In addition to his enthusiasm for science and mathematics, he showed an early interest in music, taking piano lessons from Anton Bruckner. At the age of nineteen he entered the University of Vienna, studying under Loschmidt and Josef Stefan (1835–1893), and received his doctorate after three years. Nine years later he married a woman who had begun studying mathematics at the university after meeting him but was prevented from continuing when the faculty decided to exclude women students. The marriage eventually produced five children.

Boltzmann was a sociable, restless, neurotic man endowed with a lively sense of humor and a disdain for formality, sometimes to the dismay of his more formal colleagues. When the imperial court eventually offered him a title of nobility, he refused it. He held consecutive professorships at the universities of Graz (where he became rector), Munich, Vienna, and Leipzig. Several times he traveled to the

United States, first in 1899, again in 1904 to lecture at the World's Fair in St. Louis, and in 1905 to the University of California at Berkeley as well as to Stanford. He wrote a hilarious description of his experiences in California, in a piece entitled "A German Professor's Journey into Eldorado."[3] By this time Boltzmann was suffering from deep depression, among other maladies, and in 1906, after a previous unsuccessful attempt, he committed suicide at Duino, near Trieste, by hanging himself.

Boltzmann's entire work was dedicated to the atomic theory of matter, and he regarded the kinetic theory as an essential part of atomism. At the end of the nineteenth century, however, a few prominent scientists—Ernst Mach and Friedrich Ostwald among them—accepted the kinetic theory but not the atomic constitution of matter. The disagreement centered on the "real" existence of indestructible particles, too tiny ever to be observed. Opponents based the kinetic theory on abstract "centers of energy" and saw no need for atoms or molecules. The rancorous debate between the adherents of atomism and those of "energetics" became at times quite heated, especially in Vienna, where Mach lived. Boltzmann, a strong proponent of the atomic side, was in the middle of the controversy, enduring attacks on him personally and on his work. He found more hospitable soil for his views in Britain and acquired a large reputation there before the German-speaking countries accepted his ideas. Although these attacks did not precipitate his suicide, which was undoubtedly the result of the bipolar disorder from which he had suffered for most of his life, the unpleasant atmosphere surrounding him must surely have contributed to his depression.

It is a sad irony that about a year before Boltzmann's death, and unknown to him, Einstein had published his paper on Brownian motion, which finally made the evidence for molecules palpable. And experiments performed three years after Boltzmann's death by the French physicist Jean Baptiste Perrin (1870–1942) provided convincing proof of the reality of molecules. Using particles of known mass, he was able to apply Einstein's ideas to deduce Avogadro's number from the details of his observations of their irregular movements. His result substantially agreed with the number obtained in other ways, clinching the case for atomism and vindicating Boltzmann.

What Boltzmann learned about the kinetic theory from reading Maxwell had impressed him deeply, and he immediately proceeded to build on the new statistical approach, generalizing Maxwell's distribution to hold even when forces acted on the particles. In that case, the kinetic energy in the Maxwellian distribution is simply replaced by the total energy, resulting in a probability that decreases exponentially with the energy—more slowly as the temperature rises. This energy distribution, which characterizes the state of equilibrium of any many-particle system, is still called the Boltzmann factor.

The more difficult problem was this: before it arrives there, how does such a system approach equilibrium? Maxwell had been unable to answer this question, which involves the second law of thermodynamics with its relentless, puzzling increase in entropy. Boltzmann's crucial contribution was the idea that the rising thermodynamic entropy corresponds to a growing degree of disorder at the molecular level. This increasing chaos, or randomness, and the concomitant increase in the probability of its state are responsible for the phenomenon of irreversibility that distinguishes the behavior of macroscopic objects consisting of large numbers of microscopic molecules from the reversible motions of these constituents. He promulgated an exact relation between the entropy S of a system in a given macroscopic state and the number W of the possible molecular configurations corresponding to that state, so that W is proportional to the state's probability. This relation can be written in the form of the equation $S = k \log W$ (an equation that would be engraved on his tombstone), where k is the same Boltzmann constant that appears in the Boltzmann factor and the Maxwellian velocity distribution. In order to calculate W, Boltzmann schematically considered the possible molecular energies divided up into discrete steps. The primary consideration that enters the resulting value of *W* is the number of ways in which the identical molecules of a gas could be exchanged without altering its macroscopic state, as we will see below.

For a many-particle system's development from a general state toward equilibrium, Boltzmann derived an equation governing transport processes in fluids, based on collisions between its molecules, provided that the probability of such pairwise collisions is known. This nonlinear Boltzmann equation, the first equation ever proposed to govern how a probability changes with time, still plays an important role in many different research contexts involving plasmas, neutron gases, or other fluids. Instrumental in his approach to an understanding of how the entropy of a system manages to increase while its constituents move around reversibly was a function that he invented, later called the H-function. It may be regarded as a measure of the distance between the given state of a system and its equilibrium state, in which its constituents have Maxwell-Boltzmann distributions. He then demonstrated that H obeys what subsequently became known as Boltzmann's H-theorem: the function H always decreases in the course of time unless the molecules have a Maxwellian distribution of velocities, in which case H stays constant. Here then is an explicit derivation of the irreversibility implied by the second law of thermodynamics from underlying equations of motion that are completely reversible. It also showed that the Maxwellian distribution is uniquely associated with the equilibrium state. Indeed, Boltzmann was even able to show that his H-function differs from the entropy only by a minus sign and a constant factor.

Serious objections to this curious result were raised in short order: the reversibility argument and the recurrence argument. The first, also called Loschmidt's paradox, says that, given the course of a system from some given nonequilibrium state A to a state B of equilibrium, with a corresponding increase in entropy, one need only reverse the momenta of all the molecules in state B and the system will run through its original course backward to state A, thereby decreasing its entropy. The other objection, called Zermelo's paradox after the German mathematician Ernst Zermelo (1871–1953), was based on Poincaré's recurrence theorem. Since every system must eventually return arbitrarily closely to its initial configuration, its entropy must decrease in the course of this return if it increased initially.

For the resolution of both objections it is important to remember that we are dealing with probabilities and statistics, not deterministic predictions of the behavior of individual systems. In the course of the development of a system consisting of many particles, such as a gas, the entropy varies and fluctuates, sometimes wildly. The states called equilibrium are by far the most probable because there are so many of them, all practically indistinguishable from one another. The entropy in these states is maximal. Any individual state far from equilibrium is very improbable—it has low entropy—because there are very few states like it or almost like it. Such states are always set up by external intervention; they almost never arise spontaneously in an isolated system.

Consider, for example, two adjacent rooms at very different temperatures. The second law of thermodynamics dictates that when a door is opened between them, the temperature will tend to equalize, with a concomitant rise in entropy. The initial state A of the two connected rooms of unequal temperature and low entropy could surely not be expected to arise by itself. However, it could have arisen spontaneously as an extremely rare fluctuation, taking the system from a state of high entropy to one of low entropy. A plot of an isolated system's entropy over the course of a long period, showing all its fluctuations, is symmetric in time in its general shape, showing no features that distinguish between the future and the past. If we artificially set up the system in a state of low entropy—opening the door between a hot and a cool room—then it is vastly more probable for it to develop into a state of higher entropy than to go to one of even lower entropy. The air molecules have many more states available to them in which the difference in temperature between the two rooms has decreased than states in which that difference has increased. Of course if we could freeze the situation after the two rooms have reached more or less equal temperatures (call this state S) and reverse the momenta of all the air molecules (call that state S′), the two rooms would revert back to being hot and cold, with a corresponding decrease in entropy, and the molecular state S′ will no doubt eventually be reached, at least approximately, as a spontaneous, extremely rare fluctuation, but it is for practical purposes impossible to arrange externally. Thus the reversibility objection is made harmless.

As for the recurrence argument, it too is answered by referring to fluctuations. Indeed, a system in a state of low entropy will, as Poincaré decreed, eventually return to an almost identical state of equally low entropy, but "eventually" here means that the waiting period for the recurrence to happen in an inevitable fluctuation is many times longer than the age of the universe. This is not a wild guess but can actually be calculated.

The statistical explanation of the second law of thermodynamics is also haunted by a ghost called Maxwell's demon. Of the same size, roughly, as the molecules of a gas, this little nimble-fingered fictitious imp operates a sliding door (without friction, requiring no work) on an opening between two containers filled with gases of unequal temperatures, like the two rooms above. Taking advantage of the fact that not all the molecules in a given container move at the same speed—they have Maxwellian velocity distributions—she carefully watches the particles approaching her gate from the cool side and, whenever one of them moves much faster than the average, she opens the portal and lets it pass to the hot side. In order to keep the pressures equal in the two vessels, she then allows a molecule moving particularly slowly on the hot side to pass over to the cool vessel. The result of this little game, after a while, will be to increase the temperature of the hot container and to decrease it in the cool one. In other words, the demon has managed to defeat the second law, allowing heat to flow from cold to hot without doing any work.

Much thought and ingenuity, over many years, has been expended on finding a flaw in Maxwell's demon: for example, to see the molecules, the imp needs something like a flashlight, whose electromagnetic radiation will also have to be in thermal equilibrium with the gases; and furthermore, energy has to be expended by it. The essential point, however, is that for the second law of thermodynamics to hold, there has to be an unbreachable division between the macroscopic world and its microscopic substructure. In recent years a new field of physics has opened up which operates at the borderline between the microscopic and the macroscopic; it is called mesoscopic physics. In this domain the second law of thermodynamics holds with a somewhat lower probability. The measure of this division is the large size of Avogadro's and Loschmidt's numbers. Because these numbers are so enormous, "small probabilities" of certain occurrences, in the context of the second law of thermodynamics and the associated fluctuations, are really extremely tiny. The reason is this.

From a macroscopic point of view—looking at a room full of air—there is no way of distinguishing between small differences in the way molecules are arranged. A specific point that totally escapes notice is the difference between having two molecules A and B of the same kind in places X and Y (in phase space), respectively, on one hand, and having them in places Y and X, respectively, on the other—that is, having their locations and momenta exchanged. Now, if you count the number of ways in which n particles of the same kind can be exchanged, that number starts out growing slowly with n but ends up growing extremely fast. There is a vast difference between the number of exchanges possible between the molecules of our two connected rooms when all of them are distributed á la Boltzmann at one temperature as compared to the number of exchanges possible if half of them are distributed according to one temperature and half according to another. In fact, that difference in the number of arrangements, not difficult to calculate, dwarfs the number of atoms in the entire universe. It is because of such numbers that the probability of encountering a violation of the second law of thermodynamics is so small as to be totally negligible.

Now it may appear that there is no point in making a distinction between a law that is certain to hold and one that has a completely negligible probability of ever being found violated. However, in principle the difference is revolutionary, and physicists accustomed to thinking of the second law as one of the most fundamental laws of nature, whose violation was unthinkable, had a very hard time accepting the idea that its violation was just extremely improbable. And if someone ever were to observe a spontaneous fluctuation—suddenly one of two connected rooms turning cold and the other hot, with no discernible cause—no one would believe it. The observation, being so very rare, would be impossible to repeat and hence of no scientific value. There is no denying that the introduction of probabilities in physics represented a genuine paradigm shift with some very counter-intuitive consequences, some of them not recognized until the invention of quantum mechanics and then ascribed to the weirdness of that theory. For example, the predictions of any probabilistic theory vary, depending upon whether a system was observed at some intermediate time or remained unobserved, even if we do not care what the result of the intermediate measurement was.[4]

Thus, the second revolution in physics, which replaced determinism by probabilistic predictions, began with the introduction of statistical mechanics in the second half of the nineteenth century and would come to its completion in the course of the twentieth century with quantum mechanics. Before turning to that topic, however, we will take a closer look at the meaning of the concept of probability and its history.

nine

Probability

The notion of probability originated from games of chance, and gambling is as old as history. The *talus*, the heel bone of a running animal such as deer and sheep, was found, sometimes polished and engraved, in Sumerian and Assyrian sites as well as in ancient Egyptian tombs, accompanied by score boards and other illustrations. The shape of a talus assures that it can rest on a level surface only in one of four ways. No doubt it was a precursor of the die, though a loaded one—every talus has its own bias. However, even though gambling was certainly a pastime in antiquity for many centuries (Marcus Aurelius was said to have been an avid thrower of dice), no mathematician appears to have taken an interest in it until the sixteenth century. Galileo clearly recognized that in computing the probability of a throw by several dice, what mattered was the number of permutations adding up to the same outcome, rather than the number of "partitions," as many others thought and some continued to believe.

When rolling two dice, what are the chances of an outcome of 2 as compared to 3? The only way to get 2 is for the dice to show 1 and 1; the only way of getting 3 is for them to show 1 and 2. Thus for both outcomes the number of "partitions" is the same. On the other hand, whereas the partition 1 and 1 can be obtained only one way, the partition 1 and 2 can be obtained by two permutations: die one showing 1 and die two showing 2, or die one showing 2 and die two showing 1. So counting partitions leads to equal probabilities for 2 and 3, while counting permutations leads to the conclusion that 3 is twice as probable as 2. It was not obvious at the time which was the correct calculation, and it seems that even Leibniz once made the mistake of counting partitions for probability calculations. (The difference between counting permutations and partitions would later on, in quantum mechanics, be precisely the difference between Maxwell-Boltzmann statistics and Bose-Einstein statistics. If the two dice were totally indistinguishable, as elementary particles are according to quantum mechanics, permutations would be of no account and only partitions would matter.)

How the decision in favor of permutations was finally arrived at is not entirely clear, but it was probably made on the basis of the experience of knowledgeable gamblers, that is, by observing long runs of rolling dice. Actually, it was more common in the Renaissance to play with three dice rather than two, and Galileo calculated the relative probabilities of the outcome of 12 as compared to 11. The number of partitions of each of these two outcomes is six, but the six partitions of 12 contain 25 permutations, while those of 11 contain 27. This is the reason, he says, why gamblers regard 11 as more advantageous than 12.[1]

The first book on games of chance, entitled *Liber de ludo aleae,* was written by Girolamo Cardano about 1564 but not printed until 1663. In his computations of probabilities, he defined what he meant by the numbers he calculated in terms of long series of trials, a definition that was ignored until reinvented about two hundred years later by John Venn. Indeed, the very existence of Cardano and his work was forgotten for centuries. It was Pascal who originated the modern foundation of probability theory, mostly in the course of his extensive correspondence with Fermat.

Born in 1623 in Clermont-Ferrand, France, Blaise Pascal was

brought up by his father, a civil servant but also a mathematician, as his mother died when he was three. At the age of twelve, already showing exceptional abilities, he began his scientific studies by reading Euclid's *Elements*, and by the age of sixteen he started participating in meetings of Mersenne's Académie Parisienne, held at the Convent of Place Royale, where other participants included Descartes, Fermat, and Thomas Hobbes. The death of his father induced him to adopt a more spiritual mode of life, and at 23 he converted to Roman Catholicism in the rigorous form of Jansenism. After a fervent religious experience one night at the age of 31, Pascal became quite withdrawn and wrote only at direct request from his religious adviser among the monks at the Place Royale. As his health deteriorated, he devoted himself to designing a public transport system for the city of Paris, which in fact was inaugurated during the year of 1662, when he died in Paris of a malignant stomach ulcer.

Not only was Pascal an important mathematician, but he would also acquire a wide reputation as a religious philosopher and a great writer, his fame resting principally on his *Pensées*, a book not published until eight years after his death. This volume contains his widely known wager concerning the existence of God, the underlying logic of which, as a byproduct, evolved into the art of conjecturing and would eventually form the seed for a theory of decision-making. If there is no God, you lose nothing by not believing and gain nothing by believing; but if God exists, you are saved if you believe and eternally damned if you don't believe. Hence the wiser course is to believe. Pascal was convinced that such belief was under your control: it was sure to come if you faithfully followed pious religious practice. The book also contains a sentence reputed to be the most beautiful in the French language. The sentence, in pensée no. 206, describes Pascal's reaction to viewing the night sky; it reads "Le silence éternel de ces espaces infinis m'effraie." (The eternal silence of these infinite spaces frightens me.) Generations of French school children have had to memorize and recite it.

Pascal's first significant mathematical work was based on a treatise by Gérard Desargues (1593–1662) that dealt with the intersection of a cone with a plane. As Desargues studiously avoided the use of Cartesian coordinates, his work was not understood by other mathematicians, and Pascal became his main disciple, gaining fame by creating a new branch of mathematics that would be called projective geometry. His treatise on this subject, however, was never published, and its contents became known only through reports from Leibniz, who had seen it in manuscript form.

Pascal's second project, pursued simultaneously with his work on projective geometry, was the design of a mechanical computing machine to help his father in his accounting chores. He devised a model in 1645 that served as a basis for the manufacture and sale of the first mechanical computer (seven of which still exist). Inspired by Torricelli's experiments, Pascal also made original contributions to hydrostatics, completing a treatise on the subject at the age of 31. Pascal's law governs the transmission of pressure in a fluid in equilibrium. In his honor, the international unit of stress in a fluid is called the pascal.

Among Pascal's most important work was his theory of probability (a word he never used). This theory evolved in the course of a long correspondence with Pierre de Fermat concerning two specific problems. The first was the probability of throwing a specified face of a die in a given number of successive throws, and the second was how fairly to distribute the leftover pot of stakes to the players when a game was interrupted. The correspondence developed all the needed techniques for calculating aleatory probabilities, that is, those necessary for computing odds in games of chance.

The introduction of quantitative measures for epistemic probabilities was another matter, however. Pascal's wager was an example of the application of probability in logical thinking without resorting to numerical comparisons. The first use of the concept of probability as something measurable in the context of general reasoning seems to have occurred in the book *La logique, ou l'art de penser,* written by some of Pascal's associates at the Port Royal (exactly who the authors were and how much each contributed remains under dispute). It was published first in 1662 and remained enormously popular, with translations into many languages, well into the nineteenth century.

Whereas *La logique* modeled the numerical measures of epistemic probabilities after games of chance, Leibniz's use of measurable probability stood on its own, before he even knew much mathematics. For him, probability meant degree of certainty, and he applied it largely to legal reasoning. The crucial distinction between the aleatory and the epistemic approach to probability lies in the fact that the latter is based on our knowledge of a situation rather than on its factual reality. As a result, epistemic probabilities contain a strong element of subjectivity, whereas aleatory ones may be called objective probabilities. This duality in the meaning of probability haunts the subject to this day, and as we shall see in the next chapter, it would have reverberating echoes in quantum mechanics in the twentieth century.

The notion of an average outcome, or an expected result, obvious as it seems to us now, was not at all a clearly defined concept until the later part of the seventeenth century. It was primarily the contribution of Christiaan Huygens, who equated a "fair price" for a product of uncertain value with the "expectation," that is, the average. Applying the same concept to life expectancies, he carefully distinguished between the average and the median. Suppose the individual annual incomes of five given people are \$15,000, \$16,000, \$18,000, \$20,000, and \$100,000. Their average income then is (\$15,000 + \$16,000 + \$18,000 + \$20,000 + \$100,000)/5 = \$33,800, whereas their median income is \$18,000: there are as many of them with incomes less than \$18,000 as there are with higher incomes. Ignorance of this distinction is sometimes exploited by politicians.

Words for neither of these concepts existed at the time, and it was not obvious how to use available mortality tables to answer such questions as the chance that a newborn child will die before the age of eighteen. Life-insurance premiums and the price of annuities were, of course, intimately bound up with such calculations. (Until 1789, the price of an annuity offered by the British government did not depend on the buyer's age.)[2]

The science of probability was brought to maturity by the Swiss mathematician Jacques Bernoulli, older brother of Jean Bernoulli, Daniel Bernoulli's father. Born in 1654 in Basel, Jacques was originally trained to become a theologian but studied mathematics on his own. Subsequent to a trip to England, where he met Robert Boyle, and after reading the work of Leibniz and corresponding with him, he decided to devote himself to science. In 1687 he was appointed professor of mathematics at the University of Basel, where he remained until his death in 1705. His most important papers formed the beginnings of what later became the calculus of variations, a subject that would have many applications in physics. However, his most significant achievement was his book on probability, called *Ars conjectandi,* which took him twenty years, off and on, to write and was not published until eight years after his death.

Patterned after the Port Royal *Ars cogitandi* (the Latin title of *La logique*), Bernoulli's book was extremely innovative and at the same time summed up all that was known about the subject at the time. He had a knack for relating mathematical terms, such as expectation, to everyday language, which is particularly important for the subject of probability, a field of mathematics close to common experience but in which misconceptions abound on the part of nonmathematicians. Employing vivid examples, he demonstrated, for instance, that the rule of adding probabilities holds at best for cases of independent events. Since for Bernoulli probability meant degree of certainty rather than degree of belief, he does not regard all probabilities of independent events as necessarily additive. The most important result in the book, however, is his proof of a limit theorem, now called the weak law of large numbers, establishing a relation between probability and observed frequency of occurrence. Here is what his theorem says.

Suppose we have a set-up in which at every individual trial the probability of winning is p and, running n trials in succession, we win s_n times, so that the relative frequency of winning in such a run is s_n/n. Then, the probability that s_n/n differs from p by no more than a given amount a, no matter how small, increases to 1 as n is made larger. The theorem even explains how to calculate n to make sure the probability that the difference between s_n/n and p is less than a differs from 1 by no more than another given number b. In other words, playing an honest slot machine (one on which the house takes no cut), you are essentially assured not to lose more than a penny if you play long enough, and you can even calculate how many times you have to pull the lever in order to be "morally certain" (that is, with a probability of, say, 0.999) to come out almost even.

Great as the advance was that Bernoulli's theorem made in probability theory, it did not provide the answer to a question that would preoccupy many later probabilists: while the theorem tells us the length n of a run of trials to make almost sure that the relative frequency s_n/n of winning in the run is close to the given probability p of winning in each trial, it does not tell us the length n required to make almost sure a given s_n/n is close to an unknown p. In other words, suppose you count your winnings after pulling the lever n times on a slot machine: how large does n have to be to make you confident that the probability of winning an individual game—a probability that is not given—is almost the same as the ratio of wins to pulls you have calculated from your winnings? Even though the two questions, one based on the assumption that the individual probability is given and the other on the assumption the relative frequency in a run of fixed length is given, have sometimes later been confused, they are not identical and Bernoulli provided an answer only to the first.

The other weakness of Bernoulli's theorem is this: rather than assuring that for sufficiently large n the difference between the relative frequency s_n/n and the probability p is necessarily small, it only promises that the probability of that difference being small is close to one. That still leaves open the chance that in an individual long run the relative frequency s_n/n may be quite different from p. A slight strengthening of his theorem allowed Bernoulli to find the length n of a run that would maximize the probability that s_n/n is close to a given p, no matter what the value of that p is. Thus his theorem allowed one, before performing the trials with a p that is not given, to estimate the length of a run needed in order to be morally certain the relative frequency found will be close to p. His theorem, later to be strengthened by Laplace and Poisson into what came to be called the central limit theorem of probability theory, provided the first assurance of the stability of long trial runs, an assurance on which all subsequent applications of statistics would depend. Before Laplace, however, two other, less prominent mathematicians made significant contributions to the science of probability: the religious outsiders Abraham de Moivre and Thomas Bayes.

Born in 1667 in Vitry-le-François, Champagne, Abraham de Moivre was the son of a provincial surgeon. Because he was brought up a Huguenot, he had a hard time finding suitable schooling until he moved to Paris at the age of seventeen, where he received instruction in mathematics, reading Huygens on probability as well as Euclid. Having been imprisoned for twelve months for his religious beliefs, he left for England, where he read Newton's *Principia* and became a friend—to the extent anyone could be—of Isaac Newton and of Edmund Halley. Elected to the Royal Society, he later served on the Grand Commission that tried to settle the priority dispute over the calculus between Newton and Leibniz. He eked out a living as a mathematics tutor and as a consultant to insurance companies and gambling syndicates but never obtained a permanent position. He died in London in 1754.

De Moivre made important contributions to trigonometry—an equation is named after him—but his masterpiece was his book *The Doctrine of Chances,* dedicated to Newton and republished in expanded versions both in Latin and English several times by the Royal Society. It contained the first statement of what later came to be called the normal or Gaussian distribution, a crucial element in the use of statistics (Fig. 13). The question it answered was this: in a series of trial runs with a given probability p as envisioned by Bernoulli, how were the wins and losses distributed? While the probability for a specific number of wins and losses in a run of length n can be laboriously calculated (leading to what is called a binomial distribution), the normal distribution, the now familiar bell-shaped curve, is an approximation that becomes increasingly exact as n increases and that is fairly accurate even for small values of n.

This same distribution also arises in an apparently quite different context that fascinated both Gauss and Laplace: suppose a certain quantity Q is measured by instruments of limited precision, so that every time a measurement of Q is performed, the result is not exactly

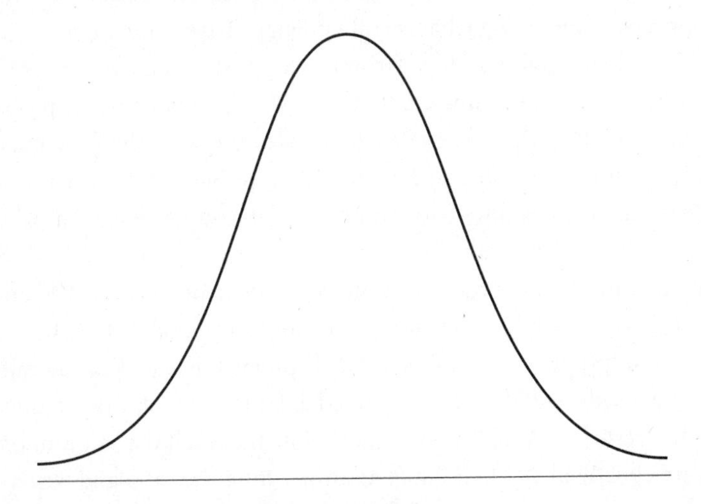

Figure 13 A normal (or Gaussian) distribution.

Q but Q_i, which differs from Q by a certain error D_i. How are the numbers Q_i distributed and how do they characterize the accuracy of these measurements? The answer is, again, that they can always be expected to be distributed "normally," and the width of the bell-shaped curve at its half-maximum can be used as an indication of the precision of the measurement, a methodology that is used in experimental physics, as well as in other sciences, to this day. Sadly, the value of de Moivre's work was not recognized until long after his death.

Now to the second of the religious mavericks. Thomas Bayes was born in London in 1702, at a time when religious dissent had just ceased to be dangerous in England. His father was one of the first ordained Nonconformist theologians. Thomas was privately educated and spent his life as a minister at the chapel in Tunbridge Wells. After publishing his book *Introduction to the Doctrine of Fluxions* (a defense of Newton's calculus against an attack on its logical foundations by Bishop Berkeley), Bayes was elected to the Royal Society; he died in 1761. His fame rests on his "Essay towards Solving a Problem in the Doctrine of Chances," published posthumously in 1763, a short essay containing what is still known as Bayes's theorem.

Here is the point of this theorem. Suppose a variety of possible mutually exclusive assumptions A_i, whose individual a priori probabilities are $P(A_i)$, enable the event E to occur with the probability $P(E|A_i)$—the probability of E on the assumption of A_i. (It is customary to call a probability based on no initial assumptions an a priori probability; on the other hand, a probability calculated by taking into account that a certain relevant event has happened is called an a posteriori probability.) Then the a posteriori probability $P(A_i|E)$ of the assumption A_i, given that E has happened, is equal to the ratio of the product $P(A_i)P(E|A_i)$ divided by the sum of the products $P(A_1)P(E|A_1) + P(A_2)P(E|A_2) + \ldots$. Note that if all the assumptions A_i are equally likely, then the $P(A_i)$ drop out of the ratio and we have a simple formula for the a posteriori probability of A_i, given that E

has happened, in terms of the probabilities for E on the basis of each of the assumptions, $P(A_i|E) = P(E|A_i)/[P(E|A_1) + P(E|A_2) + \ldots]$. For example, in a toss of three pennies, the probability that the first one shows heads, given that the total number of heads shown is two, equals the probability that the total number of heads shown is two, given that the first one shows heads, which equals ½, divided by the sum of the probability for the total number of heads shown to be two on the assumption the first one shows heads (= ½) plus the probability for the total number of heads to be two on the assumption that the first one shows tails (= ¼); the result is ⅔.

Bayes's theorem later came to be used in an extremely controversial manner for the calculation of the a posteriori probabilities of the correctness of competing theories, given that an event compatible with all of them was observed. What makes this Bayesian argument controversial is that it is based on the supposition that the a priori probabilities of the correctness of all proposed theories are the same.

Back, then, to Bernoulli's reasoning about probabilities. His basic assumption, like that of everyone else at the time, was that the individual probability p was fundamental, and it was determined by a set of outcomes each of which was equally possible either in a physical sense or else in the symmetry sense that there was no reason to think the occurrence of one of them to be more likely than any other. For a well-constructed and balanced die, thrown fairly, there was no physical cause to make a landing on one side more possible than on any other. When used in a game of chance, the notion of fairness added to the need to assume equal probabilities for all elementary outcomes.

"Equal possibility" as the basis for all probability calculations was also the starting point for Laplace, the mathematician who gave the final polish to what is now referred to as classical probability theory. Laplace had provided the epitome of the deterministic view of the universe, the kind of physics that in the nineteenth century was to be overthrown—and this same scientist also molded the underlying mathematical structure of the new thinking that was to replace determinism.

Since there could be no such thing as chance in the laws of physics, Laplace thought, the basis of the notion of chance necessarily had to be sought in our ignorance. That is why he had to choose the equal possibility of a set of outcomes as the cornerstone of a theory that dealt with instances in which we lacked full information or firm control. One such case, with which he was intensely preoccupied for a time, was the problem of induction: how much confidence could we have that the sun would rise tomorrow? This philosophical question had been raised with great skeptical effect by David Hume in his *Treatise of Human Nature,* published in 1739. Laplace tried to give a numerical answer to it, without achieving convincing success, but since it has little specific physical relevance, we shall leave it aside. Hume's criticism did, however, have great philosophical relevance for the meaning of causality, and it motivated Immanuel Kant to try to rescue science from what he saw as its devastating effects.

There were two good reasons why probability theory required modification and further development, though neither of them received serious attention until the twentieth century. One was that not all elementary natural events begging for probabilistic analysis were, in fact, equipossible. Not only did biased dice exist and made their appearance in games of chance, but the average number of girls born during a year did not usually equal the number of boys born, and these numbers were basic for the calculation of the premiums for certain insurance policies. In addition, it occurred to some imaginative thinkers to pose questions involving infinite numbers of possibilities and even continuous distributions of events rather than just finite numbers of them.

The famous problem posed in 1777 by the French naturalist George-Louis Leclerc Comte de Buffon (1707–1788) is an example of a probability with a continuous number of cases. A horizontal board is ruled by a series of equidistant parallel lines, and a fine needle, shorter than the distance between the lines, is thrown randomly on it. What is the probability that the needle will intersect one of the lines? The answer will depend on exactly what is meant by "thrown randomly." The position of the needle may be described by giving the distance of its midpoint from the nearest line and the acute angle it makes with a perpendicular dropped from this midpoint to the nearest line. Equipossibilities can then be approximated by first dividing half the distance between the lines into n equal sections and the right angle into m equal parts, postulating equal possibilities for each of these $n \times m$ cases for the position of the needle, finally making both n and m larger and larger so as to approach the continuous case. On the basis of this reasoning the probability for the needle to intersect one of the lines turns out to be $p = 2l/h\pi$ if h is the distance between the lines and l is the length of the needle. It is remarkable that the number π appears in this. Between 1849 and 1853, the astronomer R. Wolf in Zurich performed an experiment, dropping a needle 5,000 times, with a result that could be interpreted as an experimental determination of the value of π, namely 3.1596, which differs from the correct value by less than 0.02. Similar experiments performed later led to even closer results.

This way of dealing with continuously infinite numbers of possibilities, however, was mathematically rather primitive. More than a hundred years after Buffon, Poincaré raised a question reminiscent of one answered without a real mathematical basis by Nicole Oresme in the fourteenth century: what is the probability that a randomly picked number between 0 and 1 is rational? Moreover, Poincaré argued in his famous recurrence theorem that the probability for a dynamical system to start with initial conditions that would prevent it from returning arbitrarily closely to these conditions is zero. It was to deal with questions like these that more sophisticated mathematical methods were needed. And in the late nineteenth and early twentieth century, mathematicians, beginning with the Danish-born German Georg Ferdinand Ludwig Philip Cantor (1843–1918), developed such procedures. They go by the name of measure theory, and they were meant for such other purposes as set theory and newer versions of the integral calculus, but their use for the calculation of probabilities was immediately apparent and quickly exploited.

The problem was how to assign a "volume" to a given infinite set of numbers or of points in some given space, whether that set is countable or not. (A set is countable if each member can be assigned a unique positive integer. The set of all real numbers between 0 and 1 is an example of an uncountable set.) Once the notion of volume was defined, it could be used for such sets to take the place of the number of equipossible finite numbers of cases. The French mathematicians Émile (Félix-Eduard-Justin) Borel (1871–1956) and Enri Léon Lebesgue (1875–1941) and the German mathematician Felix Hausdorff (1868–1942) were the first to develop this idea in a systematic way. The lives of both Borel and Hausdorff were decisively touched by the central political events of the twentieth century. Arrested in Paris in 1940 by the Germans, Borel joined the Resistance upon his release and was awarded the Resistance Medal after the war. Hausdorff retired in 1935 as professor of mathematics at the University of Berlin under Nazi law—he was Jewish—and facing the prospect of internment in a concentration camp in 1942, he and his wife ended their lives by suicide.

Among Borel's and Lebesgue's significant contributions to the theory of functions and integration theory is Borel's definition, in 1898, of "measurable sets" of numbers and of a "measure" of such sets. (The integral calculus of Newton and Leibniz had been put on a rigorous mathematical foundation by Riemann, but the Lebesgue integral, which Lebesgue published in 1900 as one of his first papers, was applicable under much more general conditions.) He did this by generalizing the notion of the length of an interval on a line, a definition that still today is known as the Borel measure and that has the virtue of being actually computable to any desired accuracy; the Lebesgue measure, which was contained in Lebesgue's doctoral dissertation, is a powerful generalization of Borel's. Though Hausdorff's definition of measure never played as important a role in probability theory as those of Borel and Lebesgue, it turned out to be very useful in another part of physics that supplanted determinism, namely chaos theory.

The man who brought the mathematical basis of probability theory to its completion was the Russian mathematician Andreĭ Nikolaevich Kolmogorov. Born in 1903 in Tambov, where his mother had stopped on the way home from the Crimea and died giving birth to him, he was brought up by his mother's sister, Vera Yakovlevna, and her father in the village of Tunoshna, not far down the Volga from Yaroslavl, where his father was exiled, serving as a district council statistician. (He would die in 1919 in the revolutionary war.) Already by the age of seven, Kolmogorov showed great mathematical talent in school, and his aunt and adoptive mother moved to Moscow with him, where he attended the E. A. Repman private gymnasium, a school of excellent quality. Primarily interested in biology and Russian history, he studied independently while having to work with fellow senior students on railroad construction. He graduated in 1920 in the midst of revolutionary turmoil and enrolled in the Physics and Mathematics Department of Moscow University as well as in the mathematics section of the D. I. Mendeleev Institute of Chemical Engineering. Though he at first still maintained a serious interest in ancient Russian history, this was the time when he decided to devote himself entirely to mathematics.

By 1922 Kolmogorov began writing his first papers on set and integration theory, some of which attracted considerable attention, and by 1924 he turned his focus to probability theory. His first article, jointly with A. Ya. Khinchin (1894–1959), published in 1925, not only solved an important problem but also established some quite new methods that were later used repeatedly. By the time he was awarded the degree of doctor of the physical and mathematical sciences in 1935, he had developed his axiomatization of probability theory and done his fundamental work on the law of large numbers, extending and essentially completing the development begun by Bernoulli. The magnum opus, *Grundbegriffe der Wahrscheinlichkeitsrechnung* (Fundamental concepts of the calculus of probability), considered a classic today, was published in 1933 in German (the Russian edition followed in 1936).

In 1929 Kolmogorov joined the Institute of Mathematics and Mechanics at Moscow University, where he remained for the rest of his life, except for the time he spent at his dacha, a big old manor house in the woods near the village of Komarovka, where he loved to ski, walk, row, and swim in the nearby river while his aunt Vera Yakovlevna managed the household. He was promoted to professor in 1931 and appointed director of the Scientific Research Institute of Mathematics at Moscow University in 1933; eventually he would serve as dean of mechanics and mathematics from 1954 to 1958. In 1942 he married a friend from his school days.

The 1930s were an extremely productive period in Kolmogorov's life, with papers on applications of probability to biology, genetics, geology, and physics, including several papers on Brownian motion. However, around 1940 his interests began to shift toward the notoriously nettlesome problem of turbulence in the motion of fluids, to which he applied his newly evolved theory of random processes. These were the war years, though, and he also spent considerable efforts working for the military, applying statistics to the effectiveness of gunnery systems. When the war was over, Kolmogorov became the mathematics editor of the new *Great Soviet Encyclopedia,* writing 88 of its mathematics articles himself. For twelve years he served as president of the Moscow Mathematical Society and from 1982 on as editor-in-chief of the newly established journal *Uspekhi Matamaticheskikh Nauk.*

After taking up a fresh interest in information and ergodic theory, he began to attack the problem of what constitutes a random sequence. This question had become particularly acute because of the work of Richard von Mises, who advocated a definition of probability on the basis of observed frequencies in long runs of repeated trials. Other interests that preoccupied him during those years were statistics in speech and poetics, and, above all, secondary science education. He was instrumental in the founding of a boarding school, named after him and affiliated with Moscow University, in which he was personally active for some fifteen years. Kolmogorov died of Parkinson's disease in a Moscow hospital in 1987.

The mathematics of probability brought to completion by Kolmogorov, there remained the important question of its physical application and philosophical meaning. The basic Laplacian assumption of an underlying determinism in nature, still accepted in the nineteenth century even while probabilistic reasoning entered into physics and became prominent through statistical mechanics, began to crumble at the beginning of the twentieth. Consequently, simple ignorance and equipossibilities ceased to serve as the only acceptable basis for the use of probabilities, and other foundations had to be seriously explored. The French applied mathematician Antoine-Augustin Cournot (1801–1877) had already concluded that, apart from what he called philosophical probability, strict determinism was not in conflict with an objective understanding of chance, that is, probability not based on ignorance. Others who had explored this line of reasoning were the Czech mathematician Bernardus Bolzano (1781–1848), the British logician John Venn (1834–1923), as well as the American logician and philosopher Charles Sanders Peirce (1839–1914). The most important innovator in this field, however, was von Mises.

The second son of Arthur Edler von Mises, a technical official of the Austrian state railways, and his wife Adele, Richard von Mises was born in 1883 in the city of Lemberg, Austria (now Lviv, Ukraine), where his father was on temporary assignment. His parents were Jewish, but as a young man Richard converted to Catholicism, in which he later became deeply intellectually interested. After receiving a traditional classical and humanistic education at the Akademische Gymnasium, he earned his doctorate at the University of Vienna (his family's hometown) in 1907 and was appointed professor of applied mathematics at the University of Strassburg in 1909. His older brother Ludwig became an internationally prominent economist.

During the First World War Von Mises designed, supervised construction of, and personally test-flew the first large airplane for the Austro-Hungarian Flying Corps as well as filling the role of adjutant to the commander of the air force. When Strassburg became French after the war, he lost his position there and many of his possessions, and after short stints at the University of Frankfurt and the Technical University of Dresden, he was called to Berlin, where he served from 1920 to 1933 as professor and director of the Institute of Applied Mathematics at the university.

In Berlin he met Hilda Geiringer, originally from Vienna, who was his student, then his assistant and collaborator, and eventually his wife. After Hitler's rise to power, von Mises emigrated to Turkey and taught at the University of Istanbul until his immigration to the United States in 1939 to join the faculty of Harvard University. Appointed Gordon McKay Professor of Aerodynamics and Applied Mathematics in 1944, he remained at Harvard until his death in 1953.

Von Mises had a deep interest in poetry and philosophy: he was a recognized expert on the poet Rainer Maria Rilke, eventually owning the largest private Rilke collection in the world, and writing a book about him, *Bücher, Theater, Kunst,* published while von Mises was in Istanbul. He also traveled frequently between Istanbul and Vienna, where he remained a member of the Vienna Circle of logical positivists; his book *Kleines Lehrbuch des Positivismus* was published in 1939 in Holland and translated into English under the title *Positivism* in 1952.

The primary interest of von Mises as an applied mathematician was the field of hydrodynamics, on which he wrote a book while at

Strassburg and a second one, *Fluglehre* (Theory of flight), begun and first published while in the Flying Corps. However, the arenas in which he exerted his most profound influence is that of probability, on which he published two books; the first, *Wahrscheinlichkeit, Statistik und Wahrheit,* in 1928, and the second, *Wahrscheinlichkeitsrechnung,* in 1931. A third, *Mathematical Theory of Probability and Statistics,* based on his lectures in the early fifties, was published posthumously.

For von Mises, the probability of the outcome of a trial is defined by its relative occurrence in an infinite repetition of that same trial. There is no underlying definition based on equal possibilities: the probability of throwing a seven with two given dice, honest or not, is simply its relative frequency in an infinite sequence of trials, that is, the ratio of the number of times of obtaining seven to the total number of throws if the dice are thrown over and over again, infinitely many times.

The problem with this approach becomes apparent if you look back at Bernoulli's theorem. No matter how honest the dice, occasionally there will be long sequences—even infinitely long ones—of nothing but snake eyes, or of alternating fives and sixes, and so on; such runs have a probability of zero, but they do occur. Von Mises tried to deal with this problem by admitting only "random sequences" for determining a probability, but that raised another problem: how do you determine if a given sequence is "random," especially if you are able to examine only a finite part of it? Furthermore, he would admit only sequences for which the frequency of any outcome, calculated for a finite piece of the sequence, tended to a definite limit as that piece is made longer and longer, thus putting an additional restriction on what he would admit as a "collective." (Another way of putting von Mises's requirements on a collective is to rule out any possibility of a successful gambling system based on its structure.) These difficulties led many mathematicians to denigrate von Mises's frequency theory of probability or to ignore it altogether while relying entirely on the measure-theoretical approach of Kolmogorov, even though Kolmogorov himself, throughout his work, used the frequency theory of von Mises when it came to applications of probability to physical problems. Eventually, the theory of random sequences, to which Kolmogorov substantially contributed in the 1960s, led to a resolution of the difficulty (particularly, two papers of 1966 and 1971 by Per Martin-Löf, a student of Kolmogorov's at the time). The frequency theory remains the most commonly used "objective" interpretation of probability.

An opposing view, subscribed to by the British economist John Maynard Keynes (1883–1946), considers probability as a subjective quantity. The most extreme expositor of this attitude in the twentieth century—now quite divorced from any Laplacian determinism—was the Italian mathematician and philosopher Bruno de Finetti (1905–1985). For Finetti, "probability does not exist," as he stated in capital letters in the preface to his book *Theory of Probability.* Comparing the notion of an objectively existing probability to "superstitious beliefs about the existence of Phlogiston, the Cosmic Ether, Absolute Space and Time, . . . or Fairies and Witches," he considered probability "as a degree of belief a given individual has in the occurrence of a given event. Then one can show that the known theorems of the calculus of probability are necessary and sufficient conditions in order for the given person's opinions not to be intrinsically contradictory and incoherent."[3] Echoes of this view would soon be heard in some interpretations of quantum mechanics, in which physics is suffused with probability and chance.

The perceived requirement of being able to realize infinite repetitions or copies of a given system was seen by many physicists and philosophers as a real stumbling block, particularly when it came to probabilistic reasoning applied to cosmology or the universe as a whole. Karl Raimund Popper (1902–1994) made an attempt to avoid it. Born in Vienna, Popper spent most of his life at the London School of Economics and at the University of London. In 1965 he was knighted by Queen Elizabeth II. His general reputation rests primarily on his influence in the philosophy of science, particularly his denial that scientific theories are products of a systematic procedure of induction and his emphasis on falsifiability as a crucial ingredient of scientific meaning. He maintained that theories are generated as conjectures, which are subsequently subjected to experimental checks that attempt to falsify them. Once a theory has successfully weathered a sufficient number of such tests, it is provisionally accepted until either falsified or replaced by another theory of wider scope.

His ideas concerning probability were expounded primarily in his *Postscript to the Logic of Scientific Discovery,* called *Realism and the Aim of Science.* Based in principle on the frequency theory, Popper's propensity theory of probability emphasized that a von Mises collective, in order to be meaningful in defining the probability of an event, has to be generated by specific experimental conditions, such as this particular pair of dishonest dice thrown over and over again in that special manner. Therefore, the probability defined by the relative frequency in the sequence generated in this particular way may be regarded as a "propensity" of the conditions required to set it up. As a consequence, it becomes possible to speak of the probability of a singular occurrence under the specified conditions, without the need for an actual collective. On the negative side, Popper appeared to endow such propensities with physical—some would say, metaphysical—reality, like a force field. Such interpretations aside, his propensity theory is simply a reasonable variant of the frequency theory.

Now back to physics, and the completion of the probabilistic revolution that began in the nineteenth century.

ten

The Quantum Revolution

The dawn of the twentieth century saw the birth of a new era in physical science. Even as both its practical applications and the number of its practitioners grew at an enormous rate, its underlying philosophy changed drastically. The story started with heat radiation, a subject that physicists thought they understood and had just brought under control. The simplest entity on which to test this understanding was a so-called black body, an idealized object that absorbs all the electromagnetic radiation falling on it from outside without reflecting any of it back. (Such a physical system can also be realized by a hollow cavity with a small hole as its only opening.) The way the energy radiated by a black body at a given temperature is distributed over various frequencies and how that energy distribution changes with the temperature was known from experiments—for example, heating up a piece of iron until it was red-hot and noting how the color changed as the temperature increased. The German physicist Wilhelm Wien (1864–1928) had been able to show that the most intense radiation was emitted at a wavelength that was inversely proportional to the temperature; this was known as Wien's displacement law. However, no one had succeeded in justifying this law on the basis of Maxwell's theory combined with thermodynamics, a glaring failure of basic physical theory exacerbated by the fact that a detailed relation describing the emitted frequency distribution at a given temperature that Lord Rayleigh had derived strongly disagreed with experimental results. (This law, later independently derived by the British astrophysicist James Jeans [1877–1946], became known as the Rayleigh-Jeans law.) An imaginative leap was required to rescue physics, and this leap, which led to much intellectual ferment, was taken not by a young revolutionary but by an already well-established German physicist, Max Planck.

Max Planck was born in Kiel, Germany, in 1858, the sixth child of a professor of jurisprudence at the University of Kiel. When he was eight years old, the family moved to Munich, where he was educated at the Maximilian Gymnasium and subsequently at the University of Munich. After studying mathematics and physics with Gustav Kirchhoff and Hermann Helmholtz in Berlin, he obtained his doctorate at Munich in 1879 with a dissertation on thermodynamics. His first academic position was at the University of Kiel, after which he moved to the University of Berlin, where he became professor of physics in 1892, a position he retained until his retirement in 1926. Married twice—his first wife died in 1909—Planck had four children. His eldest son died as a soldier in the First World War, both of his twin daughters died shortly thereafter, during childbirth, and his second son was executed in 1944, accused of being involved in a conspiracy against Hitler.

A religious and politically conservative man, Max Planck was one of the signatories of a notorious manifesto "of ninety-three intellectuals" issued in 1914 that disclaimed any German responsibility for the war and denied that the widely reported alleged misconduct by its army in Belgium could possibly have occurred, but he soon bitterly regretted his rash signature. This lapse notwithstanding, for the entire first half of the twentieth century Max Planck would represent what was best in German science at a time when remaining honorable was difficult and even dangerous in his country. He was instrumental in persuading Einstein to come to Berlin in 1914, creating a special position for him there. Appointed president of the Kaiser Wilhelm Institute in Berlin in 1930, he resigned from this post in 1937 in protest against the treatment of Jewish scientists by the Nazi government. After the Second World War, the Institute was renamed the Max Planck Institute and moved to Göttingen, with Planck reinstated as its president. He died in Göttingen in 1947.

Scientifically, Max Planck embodied the reluctant but willing transition of the old guard from classical physics to an entirely new probabilistic paradigm in which iron-clad laws changed to statistical regularities. To admit the possibility, even if rare, of a violation of the second law of thermodynamics by a fluctuation was a wrenching thought for him, and he only grudgingly accepted atoms, which, after all, were ultimately responsible for the statistical nature of thermodynamics. Yet he was the physicist who fired the first shot in a scientific revolution that he did not welcome. In 1900, without any underlying rational justification other than strictly mathematical purposes, he applied Boltzmann's probabilistic reasoning to calculate the entropy of the energy distribution in a black body by means of an unheard-of assumption: that the energy E of electromagnetic radiation of a given frequency f emitted or absorbed by the walls of a cavity could not take on any arbitrary value but had to be proportional to that frequency, $E = hf$, or integer multiples thereof. The constant of proportionality, h, now considered one of the most fundamental in nature, came to be known as Planck's constant. (Actually, Boltzmann himself had used discrete energies for his calculation of probabilities, but his discretization was not tied to a radiation frequency as Planck's was.) In that way—and only that way—Planck was able to obtain Wien's displacement law, and eventually even to derive a new energy-distribution law for black bodies to take the place of the Rayleigh-Jeans formula. Planck's new radiation law turned out to agree well with experimental data. This would not be the last time in this field that imaginative scientists generated productive new theoretical ideas that appeared totally unjustified.

The next step in the revolution was taken by Einstein in 1905, but to understand it we have to retrace our steps a bit. In the course of their research on electrical conduction through gases, the German physicists Julius Plücker (1801–1868) and Johann Wilhelm Hittorf (1824–1914) had found a new kind of ray, now called cathode rays, whose presence becomes visible by a glow either in a gas or on the surface of a glass vessel containing a gas (such as the CTR monitor of a computer or television picture tube). The nature of these mysterious rays, which could be deflected by a magnet—now known to be streams of electrically charged particles called electrons—was uncovered in 1897 by the British physicist J. J. Thomson (1856–1940). (This discovery was not unique to Thomson. The two German physicists Emil Wiechert [1861–1928] and Walter Kaufmann [1871–1947] both also independently discovered the nature of cathode rays at about the same time.)[1] In the meantime, Heinrich Hertz had found in 1887 that electric spark discharges were enhanced by exposure to bright light. What he had discovered came to be known as the photoelectric effect. In terms of J. J. Thomson's discovery, Hertz had found that when light shines on a metal surface, the surface emits electrons. Some fifteen years later, when Philipp Lenard investigated the photoelectric phenomenon in more detail, his data showed something very puzzling about this emission: increasing the brightness of the light produced more electrons, but it did not increase their speed, which was entirely determined by the color of the light. No one was able to understand this puzzle until Einstein solved it as part of one of his three groundbreaking papers in the year 1905.

If light of frequency f consisted of particles, later called photons, each of which had the energy $E = hf$ (the same connection between energy and frequency that Planck had used in a somewhat different context, and with the same constant h), then Lenard's observation was easily explained: the ejection of each electron was the result of a collision with a photon; the brighter the light, the more photons, and hence more electrons; blue light has a higher frequency than red, so its more energetic photons will give the electrons a bigger kick than red light will.

Did this mean that Einstein denied the validity of Thomas Young's interference experiment and wanted to return to Newton's corpuscular theory of light? No: in the first of many counter-intuitive aspects of the quantum theory, light, somehow, behaved both as a wave of frequency f *and* as a particle of sorts of energy $E = hf$ (though photons could not be just "little billiard balls"). Thus was born the quantum theory of light. The most direct confirmation of the particle nature of light came in 1923, when the American physicist Arthur Holly Compton (1892–1962) discovered what came to be known as the Compton effect: when electromagnetic waves, such as X-rays, are scattered by electrons, their wavelengths are changed exactly as if they were particles colliding at an angle, bouncing off with reduced energy (consequently with reduced frequency, by Planck's relation, therefore with increased wavelength, since the wavelength is inversely proportional to the frequency).

In the last decade of the nineteenth century all sorts of new rays were found, and the unraveling of their nature added unexpected aspects to the still controversial theory of atoms. First came the discovery of X-rays in 1895 by the German physicist Wilhelm Konrad Röntgen (1845–1923). In the course of studying the properties of cathode rays, which caused luminescence in certain chemicals, he found that the glow continued to be produced when the cathode-ray tube was covered by cardboard, and even when the chemical was taken into the next room. The penetrating rays were obviously not the same as cathode rays but were produced as cathode rays hit the glass walls of the tube. He named these previously unknown, mysterious rays X-rays. When he went public with his discovery, showing photographs of his hand with its bone structure made visible because the X-rays easily penetrated skin and tissues but were absorbed by bones, he caused a sensation. In 1912, when the German physicist Max von Laue (1879–1960) sent X-rays through crystals, he found a diffraction pattern similar to the one Thomas Young had found when shining light through slits. He concluded from this experiment that X-rays (which could not be deflected by electric or magnetic fields) were electromagnetic radiation like light but of much shorter wavelength. Von Laue's patterns also demonstrated that crystals were made up of regular arrangements of atoms and that the observed fringes gave detailed information about the structure of the diffracting crystals—a fact established by the young British physicist William Lawrence Bragg (1890–1971), together with his father William Henry Bragg (1862–1942). Some forty years later, this insight would help in the unraveling of the structure of DNA by Francis Crick and James Watson.

During the same year, 1896, that Röntgen announced his discovery of X-rays, the French physicist Henri Becquerel (1852–1908) quite accidentally found radioactivity. In his investigation to determine if the fluorescence of certain crystals when exposed to light produced Röntgen's X-rays, he had left some uranium salts wrapped in paper on a photographic plate in a drawer. Since the salts were in the dark, they could not fluoresce. To his astonishment he later found that the plate was clouded, which indicated that the uranium salt must have emitted a new kind of radiation that penetrated the paper and clouded the plate. His subsequent experiments revealed that the emitted radiation, later called beta rays, could be deflected by a magnet and seemed to have the same properties as the particle rays J. J. Thomson had found: they consisted of electrons. Becquerel's serendipitous finding, however, was only the first step in the discovery of radioactivity; the real heroine of that story was Marie Curie.

Born in 1867 in Warsaw, Poland, at that time under Russian domination, Maria Sklodowska was the fifth child of parents who were both teachers. Her father taught mathematics and physics at a secondary school and her mother managed a private boarding school for girls. When Manya, as she was called, was ten, her mother died. As her father was kept from decent teaching positions for political reasons, the family had to make ends meet by taking in boarders. After graduating from high school with a gold medal, Manya took a job as a governess, sending part of her earnings to support her older sister Bronia, who was studying medicine in Paris. In addition to a passionate involvement in Polish political causes, Manya voraciously read not only science but all the literature she could lay her hands on, from Dostoevsky to Karl Marx, and gave lessons to the peasant children on the estate of her employers. Following her sister to Paris at the age of 23, she managed to obtain a scholarship and quickly passed her *licence* in physics and mathematics, both with high honors.

Three years later Manya met the physicist Pierre Curie, eight years older than she and employed as director of laboratory work at the École Municipale de Physique et Chimie. Pierre had been born in 1859 in Paris, the son of a physician, educated at home and subsequently at the Sorbonne. His experimental research made several lasting contributions to our knowledge of the way magnetic properties of materials change with temperature. In 1895 Manya and Pierre were married, and after two years Manya gave birth to their daughter Irène, who would grow up to become a famous physicist in her own right.

From 1896 on, Marie Curie became intensely interested in the new discoveries by Röntgen and especially in the rays given off by uranium that Becquerel had found. What she wanted to investigate, together with Pierre, who abandoned his own work to join her, was whether there might be other substances emitting such radiation. After finding that thorium did too, and that the mineral pitchblende emitted much more radiation than could be accounted for by its uranium content, they discovered two new radioactive elements, polonium and radium. All these substances produced other radioactive elements as well as either helium or beta rays. All of this research was done in a substandard laboratory in the École de Physique et Chimie, which was the only place of work available to them. Many honors followed, but it was not until the year after their joint Nobel Prize in physics with Becquerel in 1903 that Marie was appointed as Pierre's assistant with a salary—she had been working without pay until then. In 1904 their second daughter, Eve, was born, and Pierre was appointed to a newly created chair in physics at the Sorbonne.

Two years later, after Pierre Curie was accidentally killed by a horse-drawn carriage, Marie was appointed to the chair her husband had occupied, the first woman to teach at the Sorbonne. In 1910 she declined the Legion of Honor, as had Pierre in 1903, and she never accepted any royalties for the lucrative industrial applications of radium. The following year Marie Curie was awarded the Nobel Prize in chemistry for isolating polonium and radium as pure metals; it was the first time that a person had won two Nobel Prizes in science, and would remain the only time for more than half a century.

During the First World War, Marie Curie aided the French army's care for the wounded by helping to equip ambulances with X-ray apparatus, and the International Red Cross appointed her head of its Radiological Service. Assisted by her daughter Irène, she also created new courses in radiology for medical orderlies. Impressed by Madame Curie's generosity and scientific accomplishments, American women made her a gift of a gram of radium, paid for by national subscription and presented to her in 1921 by President Harding while she was on a tour of the United States. By the late 1920s, however, her health began to deteriorate; she underwent four cataract operations and developed lesions on her fingers, the result of handling radium without proper protection. A few months after Irène (1897–1956) and her husband Frédéric Joliot (1900–1958) announced their discovery of artificial radioactivity, Marie Curie entered a nursing home and then a sanatorium in Sancellemoz in the French Alps, where she died in 1934.

However important the discovery of radioactivity as a fact of nature was, its long-range significance consisted in raising basic questions: Where did the electrons and the helium originate? Why were new radioactive elements produced as byproducts? Rather than being the stable ultimate building blocks of nature, were atoms themselves changeable or unstable? Did the ejection of energetic particles not imply that conservation of energy was violated? These were fundamental questions, and the first physicist to offer some answers while at the same time adding immensely to our knowledge of atoms was Ernest Rutherford.

Born in 1871 near Nelson, New Zealand, his father a farmer and wheelwright, Rutherford loved to build models and showed himself to be a very bright boy, but he exhibited no special scientific aptitude as a child. After entering Nelson College (what we would call a prep school) at the age of sixteen, he went on to Canterbury College, Christchurch, receiving his B.A. degree at 21, and began studying physics and mathematics, obtaining B.S. and M.A. degrees. In 1895 he sailed for England to study in Cambridge at the Cavendish Laboratory, as J. J. Thomson's first research student. Three years later, he left for Montreal, where he had been offered a professorship in physics at McGill University. The laboratory facilities there, at that time superior to any in Britain, enabled him to perform experiments on radioactivity. In 1900 he returned to New Zealand to marry Mary Newton, with whom he had fallen in love while in Christchurch. They had one daughter, born in 1901. Leaving Montreal for England in 1907, he accepted the chair in physics at Manchester University, where he built a world-famous laboratory, second in England only to Thomson's Cavendish, and studied the atom.

After working on methods for locating submarines during the First World War, Rutherford moved to Cambridge as professor of physics and director of the Cavendish Laboratory, succeeding Thomson, and was appointed professor of natural philosophy at the Royal Institution as well. He was knighted in 1914, given a peerage in

1921 (taking the title Lord Rutherford of Nelson), and was personally presented with the Order of Merit by King George V. From 1925 to 1930 he served as president of the Royal Society, and during the 1930s he was president of the Academic Assistance Council, helping refugee scientists who had escaped Nazi Germany. Rutherford, a man who "had no cleverness—just greatness," died in Cambridge in 1937, after an accidental fall while gardening.[2] The greatest experimental physicist of the twentieth century was buried in Westminster Abbey.

A few weeks after Röntgen discovered X-rays, Thomson put his new research student to work on measuring the effects of these rays on electric discharges in gases. Rutherford found that they produced carriers of positive and negative electricity, the negative ones shortly to be identified by Thomson as identical to his electrons. When radioactivity was discovered by Becquerel in uranium and by the Curies in other elements such as radium, Rutherford turned his attention to the emissions in these processes. He found that there were two quite disparate kinds, one easily absorbed and the other more penetrating; he named them alpha and beta rays. The former would later be shown to be doubly ionized helium atoms—in other words, helium atoms carrying a double positive electric charge—and the latter, electrons. (Rutherford made this identification after moving to Manchester and after the chemists Frederick Soddy [1877–1956, English] and William Ramsay [1852–1916, Scottish] had ascertained that indeed helium was produced in the course of the transmutation of radium.) However, with Soddy's extremely valuable collaboration in these experiments at McGill, he also found that some of the emanations, as he called them, of radioactive elements produced radioactivity themselves, giving rise to other radioactive daughters until the chain ended with a stable element; in the case of radium, the end product was lead. The activity of most of these daughter elements decreased according to an exponential curve, each with a characteristic half-life (the time during which an individual atom has a 50/50 probability of decaying, or in statistical terms, the time after which half of any given large number of radioactive atoms have decayed). Exactly how the helium atoms and electrons could emerge from the radioactive atoms was still a mystery.

In 1902 Rutherford and Soddy announced a rather surprising theory they called transmutation. In the course of radioactivity, they said, elements are transformed into one another; it is both an atomic phenomenon and a chemical one, in which new kinds of matter are produced, as though alchemy had been resurrected. The reason why the activities of some elements such as radium and thorium seemed to remain constant and never diminish—thus appearing to create energy out of nothing—was simply that their half-lives were extremely long. Since these activities nevertheless tended to disappear eventually, together with the elements that caused them, radioactivity did not violate the conservation of energy, as some had feared. The new theory of radioactivity was almost instantly accepted, aside from a few conservative skeptics like the elderly Lord Kelvin.

When he came to Manchester, Rutherford focused on experiments with his alpha particles, the positively charged helium ions, which, he found, produced visible scintillations when striking a luminescent screen. Later, Hans Geiger, the research assistant who had helped with these experiments, invented a more convenient method of detecting charged particles, which is still known as the Geiger counter. In 1909, Rutherford and Geiger suggested to Ernest Marsden, an undergraduate at the time, that he investigate the scattering of alpha particles as they pass through a thin metal foil, paying particular attention to large-angle scattering. To their surprise, Marsden, with Geiger's help, found that a small but significant number of alphas were deflected by more than ninety degrees, that is, in the backward direction! "It was almost as incredible as if you fired a fifteen-inch shell at a piece of tissue paper and it came back and hit you," was Rutherford's reaction. The model of the atom that Kelvin and J. J. Thomson had proposed to account for the radioactive emission of electrons—the negatively charged electrons distributed like plums in a positively charged pudding, leaving the atoms neutral—could not possibly account for these experimental results.

By 1910 Rutherford came to the conclusion that the only way such backward deflection of alpha particles, much heavier than electrons, could occur was for the positive electric charge in an atom, and most of its mass, to be concentrated in a central region whose diameter was only about a 100,000th of that of the atom, the electrons somehow surrounding this nucleus and leaving most of the atom empty space. Rutherford had discovered the nuclear atom, and his discovery would occupy physicists intensely for the next half century, with repercussions ranging from geology to astrophysics to world politics.

But at first, the scientific community could not have cared less. Then in 1912 a young Danish theoretical physicist who joined Rutherford's laboratory not only paid attention but used and extended the new knowledge to brilliant advantage. His name was Niels Bohr. It was clear to him that radioactivity had to be an action of the central nucleus, whereas chemical phenomena were the action of the electrons in the atom.

With the help of Bohr's intuitive theoretical understanding, Rutherford's discoveries of the nucleus and of the regularity properties of radioactivity—its intrinsic randomness notwithstanding—would pave the way for an eventual physical basis of the periodic table of the elements, which the Russian chemist Dmitri Ivanovich Mendeleyev (1834–1907) had introduced with great effect in chemistry but which was still regarded by many as no more than a fortuitous book-keeping device. Rutherford's discoveries would also usher in the age of atomic and nuclear physics, especially after he went on, a short time after World War I ended, to discover artificial transmutations of nuclei, in which the collision of an alpha particle with a stable nucleus knocks off a proton—a hydrogen nucleus—thereby bringing forth a new element. In 1934 Rutherford even achieved for the first time what later would be called fusion; together with

Paul Harteck (1902–1985, German) and Marcus Oliphant (1901–2000, Australian), he produced helium as well as tritium by using the newly discovered heavy water to bombard deuterium with deuterons. (In modern language, a deuteron is a nuclear particle made up of a proton and a neutron; deuterium is the analogue of hydrogen, but with a deuteron as its nucleus rather than a proton; heavy water is a compound like water, except that deuterium replaces hydrogen. Tritium is the analogue of hydrogen, but with a triton as its nucleus; a triton is made up of a proton and two neutrons.) In other words, two lighter atoms were fused into one heavier one. Fundamental as atoms were, they could no longer be regarded as the ultimate, stable, and unalterable constituents of matter.

Niels Bohr was born in 1885 in Copenhagen. His father was a professor of physiology at the University of Copenhagen and his mother the daughter of a Jewish banker. His younger brother, Harald, probably the more brilliant of the two boys, grew up to become a well-known mathematician. After showing himself in school to be talented but not outstanding, Niels entered the University of Copenhagen to study physics, mathematics, and chemistry, earning his doctorate at the age of 25 with a dissertation on the behavior of electrons in metals. Leaving immediately for Cambridge but finding Thomson indifferent to his ideas, he moved on to Manchester, the domain of Rutherford, who had just discovered the nuclear atom.

The next year, 1912, Bohr returned to Copenhagen, unsuccessfully applied for a professorship in physics at the university, and married Margrethe Nørlund. Between the years 1916 and 1928 they had six sons. Christian, the eldest, died tragically at the age of seventeen in a sailing accident, his father on board but unable to help. After returning to Manchester, where Bohr became famous for his model of the atom, he prevailed upon the Danish government to appoint him to a professorship at the University of Copenhagen. The government also built the Institute for Theoretical Physics for him, of which he became the director in 1920, a position he retained until his death in 1962.

Bohr's institute became the center for the development of quantum mechanics, at its height attracting all the major players from around the world at least for a temporary stay. After the ascent of Hitler, he joined the governing board of the Danish committee for the support of refugee intellectuals. During the Second World War, when the German army occupied Denmark, Bohr was active in the resistance movement. But being in special danger because of his Jewish mother, he escaped to Sweden in a fishing boat, and from there flew to England. He then sailed to the United States, where he participated in the development of the atomic bomb at Los Alamos (retaining the code name Nicholas Baker he had used during his escape from German-occupied Denmark). After the war, Bohr became a strong advocate for international control of nuclear weapons, writing a famous open letter to the United Nations in which he pleaded for a free exchange of ideas among all nations of the world. In 1952 he was instrumental in the establishment of CERN, the European Centre for Nuclear Research at Geneva. Bohr died of heart failure in Copenhagen in 1962.

When Bohr joined Rutherford's Manchester laboratory, he made it his primary task to try to understand how the nuclear structure of the atom could possibly remain stable. The particle nature of electricity, that is, the electric charge of the electron, was finally pinned down in 1913 by the American physicist Robert Andrews Millikan (1868–1953) by means of an accurate measurement known as the Millikan oil-drop experiment, repeated with some difficulty by physics students to this day the world over. But the Maxwell equations decreed that an atomic structure such as the one proposed by Rutherford would quickly collapse, since electrons orbiting a central charge (like planets in the solar system) would lose their energy by radiating electromagnetic waves as they spiraled toward the center. Taking his cue from Planck, Bohr postulated as an ad hoc solution that an electron in the atomic solar system would remain stable only in certain specific orbits, namely those for which its angular momentum is an integral multiple of Planck's constant *h*. Moreover, when an electron makes a jump from an orbit of energy E_1 down to one of lower energy E_2, it emits electromagnetic radiation whose frequency f is connected to the energy loss $E = E_1 - E_2$ of the atom by Planck's relation $E = hf$. (Conversely, an atom of energy E_2 on which radiation of that same frequency shines will absorb it and jump to the energy E_1.) In the orbit of lowest energy, called its ground state, the atom remains stable forever. None of these rules was based on, or even compatible with, any known physical laws. But they could easily be applied to the simplest atom, that of hydrogen, which contained only one electron, and compared to the spectrum of light emitted by heated hydrogen gas, whose atoms were "excited" by collisions.

Every element, when heated up, was known to emit light consisting of a mixture of characteristic colors or frequencies called its spectrum. When passed through a prism, as in Newton's experiments, these colors are separated and, on a photographic plate, form a series of spectral lines that identify the element like a fingerprint. The science of spectroscopy had been initiated by the German physicists Joseph von Fraunhofer (1787–1826) and Gustav Robert Kirchhoff (1824–1887). Fraunhofer also discovered dark lines in the spectrum of the sun, which were later understood to be caused by the absorption of light rather than emission. Perhaps the most spectacular result of spectroscopy was the discovery of the element helium in the sun before it was known on the earth. The spectrum of hydrogen had been handily expressed in 1885 by the Swiss school teacher Johann Jacob Balmer (1825–1898) in a simple formula, and Bohr's rules for the energy levels led exactly to Balmer's formula for the spectrum of the emitted radiation. Justified or not, the Bohr atom worked. Indeed, it was further confirmed in 1914 by an experiment carried out by the German-born American physicist James Franck (1882–1964) together with Heinrich Hertz. Directing a beam of electrons at a bulb of mercury vapor, they observed that at certain specific energies the electrons experienced "resonances" and the mercury vapor began to glow. What was happening, they realized, was that some beam electrons were colliding with electrons in the mercury atoms, kicking them up to an excited level, subsequent to which these atomic electrons jumped back down, causing the emission of light, all in accordance with Bohr's picture.

At this point, Einstein again entered the picture by making a direct connection between Bohr's quantum jumps and Planck's explanation of the spectral distribution of black-body radiation. Prompted by the observed randomness and unpredictability of individual radioactive emission, he introduced the idea of probability—forever after the prime characteristic associated with the quantum theory—to explain the absorption and emission of radiation by atoms and molecules in one of Bohr's quantum states. This transition probability, translated into statistics for large numbers of atoms, would account for the variations in intensity of light emitted by elements in different atomic transitions. Bohr's discovery of the atomic emission of radiation of a specific frequency was thereby directly connected to Planck's hypothesis in the black-body law. Though this did not endow either of these notions with any more physical justification, at least it reduced the number of postulates that had to be swallowed ad hoc.

Calculating the probability of emission of photons by atoms, Einstein distinguished between spontaneous transitions and stimulated ones. The latter occurred when electromagnetic radiation of the proper frequency shined on an atom in an excited state, inducing it to jump to a lower state and emit radiation of the same frequency on its own. Some forty years later, the American physicist Charles Townes (b. 1915) and independently the Russians Nikolai Gennadiyenich Basov (1922–2001) and Aleksandr Mikhailovich Prokhorov (1916–2002) would experimentally exploit this process in their invention of the maser (acronym for "microwave amplification by stimulated emission of radiation") in 1953, which was followed up with the first construction of a laser ("light amplification by stimulated emission of radiation") by Townes together with Arthur Leonard Schawlow (1921–1999) and independently shortly thereafter by the American physicist Theodore Maiman (b. 1927). After the Dutch-born American Nicolaas Bloembergen (b. 1920) began to develop the experimental area of laser spectroscopy, the new fields of optical pumping and quantum optics turned out to have enormous practical repercussions in many areas of physics and technology. The theoretical underpinning of these fields was provided, respectively, by the French physicist Alfred Kastler (1902–1984) and the American physicist Roy J. Glauber (b. 1925).

Forming the beginning of what later came to be known as the "old" quantum theory, Bohr's rules were soon generalized by the German physicist Arnold Sommerfeld (1868–1951) and by Einstein to make them applicable to mechanical systems other than atoms, but the resulting conglomeration of ad hoc rules could hardly be called a theory. In addition to its reliance on probabilities, its central feature, well confirmed by experiments, was that physical systems which previously were thought to have a continuous range of possible energies were now allowed only specific discrete values—just as Planck had anticipated in order to account for black-body radiation, and Einstein had for light—but it still lacked any kind of internal coherence. Nevertheless, the Bohr model of the atom explained many intricate aspects of the spectra of light emitted by elements, and it also accounted for the regularities of the periodic table of the elements, except for a puzzling feature: in order to build up an atom of a heavy element in its stable ground state, its many electrons all had to be assumed to move in different orbits. Why did they not all congregate in the level with the lowest energy? The puzzle was solved by Wolfgang Pauli.

The son of a distinguished professor of chemistry, Pauli was born in 1900 in Vienna. While in high school, he filled periods of boredom reading the history of classical antiquity, advanced works in mathematics, and Einstein's papers on the general theory of relativity. To study theoretical physics, he went to Munich and worked with Arnold Sommerfeld, who gave him the task of writing an article about Einstein's theory for the newly published *Encyklopädie der mathematischen Wissenschaften*. The result was a 250-page monograph, written with such critical acumen and clarity by the nineteen-year-old Pauli that it remained in print as a classic for many years.

For the rest of his life, Pauli would be the super-critic whose judgment about new ideas, avidly sought and often acidly expressed, was extremely acute and almost always correct—with a few glaring exceptions. (An example will be given later.) He was also so clumsy in the laboratory that if an experiment failed within a mile of him, he was jokingly blamed. After receiving his doctorate in 1922 he became Max Born's assistant in Göttingen, absorbed the wisdom of Bohr in Copenhagen, moved in 1923 to the University of Hamburg, and then in 1928 moved to Zurich, succeeding Peter Debye upon his retirement as professor of theoretical physics at the Eidgenössische Technische Hochschule (ETH). The following year he married a young dancer, Käthe Deppner, who soon left him, driving Pauli into psychoanalysis with a man he greatly admired, Carl Jung. His subsequent marriage to Francisca Bertram in 1934 turned out to be stable. He remained at the ETH until the end of his life, except for a stay at the Institute for Advanced Study in Princeton from 1940 to 1945. Upon his return he became a Swiss citizen, and he died in Zurich in 1958.

Apart from his extremely constructive role as a critic, Pauli made two important contributions to fundamental physics. The first was his improvement on the Bohr atom. With no more basic physical justification than Bohr's idea that the electrons' orbits had to be defined by certain specific quantum numbers, he postulated what came to be known as Pauli's exclusion principle: no two electrons were permitted to occupy an orbit with the same quantum numbers. This principle at once produced the ladder of atoms whose electrons moved in larger and larger orbits—rather than all squirming into the lowest one—so as to account for the regularities of the periodic table. Later it would play an extremely important role in explaining the structure of matter and the constitution of atomic nuclei.

As an incidental byproduct of this suggestion, he also postulated that a fourth quantum number, with only two possible values, be added to the three previously used (two quantum numbers specified the magnitude and z-projection of the angular momentum, and the third, called principal, specified the electron's order in the hierarchy). He thereby clarified some of the earlier spectral formulas that did not quite work. His two-valued quantum number, for the manipulation of which he later introduced a very useful mathematical technique, would soon turn out to be physically explained by a novel property of the electron: its spin angular momentum, discovered in 1925 by two young Dutch doctoral students at the University of Leiden, Samuel Goudsmit (1902–1978) and George Uhlenbeck (1900–1988).[3]

Pauli's second fundamental contribution was his explanation in 1930 of a puzzling apparent loss of energy and angular momentum during the process of radioactivity in which an electron was emitted (also called beta decay). Whereas Bohr was ready, in desperation, to call the sacred conservation laws into question, Pauli suggested that the culprit was a massless—or almost massless—electrically neutral particle emitted at the same time, carrying away both some kinetic energy and half a unit of angular momentum. Enrico Fermi later named this elusive particle, which interacted only extremely weakly with any others and hence was very hard to detect directly, the neutrino. Although indirect evidence for its existence accumulated rapidly, almost thirty years passed before it was found. The neutrino was experimentally detected in 1959 by Frederick Reines (1918–1998), Clyde L. Cowan (1919–1974), and their collaborators F. B. Harrison, H. W. Kruse, and A. D. MacGuire. (Today, we say that a neutrino accompanies the emission of a positron, and an antineutrino that of an electron.)

For the production of a much more general theory than the Bohr-Sommerfeld "old quantum theory," eventually called quantum mechanics, with all the needed coherence and comprehensiveness, we have to thank three physicists: Heisenberg, Schrödinger, and Dirac, as well as Max Born, but Niels Bohr also played an important part. Werner Karl Heisenberg, born in 1901 in Würzburg, Germany, was the son of a very scholarly father who taught ancient languages at a Gymnasium (later appointed professor of Greek philology at the University of Munich). Beginning piano lessons at an early age, Heisenberg became an excellent player, an accomplishment he retained throughout his life. In school he showed an outstanding mathematical talent and intended to study pure mathematics when he entered the University of Munich, but instead he wrote his doctoral dissertation under the guidance of the mathematical physicist Arnold Sommerfeld, obtaining his Ph.D. in 1923.

These were years of political turmoil in postwar Germany, and Heisenberg shared the romantic ideals of a nationalistic German youth movement in which he became a leader. But in science his most profound inspiration came from Niels Bohr, whose lectures he had occasion to attend in 1922. For the next three years Heisenberg served as assistant to Max Born in Göttingen, with several extended visits to Munich and Copenhagen to work with Bohr, whose assistant he became in 1926. In 1927 he accepted a professorship in theoretical physics at the University of Leipzig, where he remained until his appointment as director of the Max Planck Institute in Berlin in 1942. He married Elisabeth Schumacher in 1937, and they had seven children.

After Hitler came to power, Heisenberg verbally defended Jewish and leftist physicists against the Nazi government, without notable success; a German nationalist but no Nazi, he refused to leave Germany in protest. The leaders of the "German physics" movement,

Johannes Stark (1874–1957) and Philipp Lenard, and their followers, attacked him for representing "Jewish physics"—this meant theoretical physics, with Albert Einstein their arch-villain. Heisenberg's precise role in Germany's unsuccessful endeavor, which he headed, to develop nuclear energy and possibly a nuclear bomb during the Second World War remains controversial. After the war, he became director of the Max Planck Institute in Göttingen, which in 1958 was moved to Munich, and he remained its director until 1970. Heisenberg died in 1976 in Munich.

Born in 1887 in Vienna, Erwin Schrödinger was the oldest of the three founders of quantum mechanics, barely two years younger than Bohr. He was the only child in an educated, upper middle-class family, brought up as a Protestant in a mostly Catholic country. After elementary home-schooling, he attended the Akademisches Gymnasium, which offered a traditional curriculum in the humanities, and excelled especially in the ancient grammars, as well as in mathematics and physics. Enrolled in the University of Vienna, he began studying theoretical physics as soon as it was offered again after the death of Boltzmann, but he also regularly attended the theater, taking notes on the backs of the printed programs. At the age of 22 he received his doctorate and remained at the university as an assistant until the outbreak of the First World War four years later.

During the war, he served as an artillery officer at a post where he could pass his time immersed in the philosophy of Spinoza and Schopenhauer when he was not reading physics, and particularly learning about Einstein's general theory of relativity. The war over and the Austrian monarchy destroyed, the only job immediately available to him was his old assistantship back in Vienna. He also met Annemarie Bertel, whom he married in 1920. Academic positions in Germany being easier to come by, he accepted a number of short-term appointments in Jena, Stuttgart, and Breslau before moving to Switzerland to become professor of physics at the University of Zurich in 1921. In 1927 Schrödinger was offered the chair at the University of Berlin vacated by Max Planck at his retirement, which he accepted.

Outraged at the dismissal of outstanding Jewish scientists and others by the Hitler government, Schrödinger left Germany in 1933, becoming a fellow at Magdalen College, Oxford. By 1936 he felt so homesick that he returned to Austria, accepting a position at the University of Graz. Germany's annexation of Austria put an end to this homecoming, when he was subjected to retaliation for his earlier flight from Germany and dismissed from his post. Coming to his rescue, Eamon de Valera, the prime minister of Ireland at the time, arranged for him to leave Austria, to pass through England after the outbreak of the Second World War on his way to Ireland, and secured a position for him at the newly established Dublin Institute for Advanced Studies. There Schrödinger remained for the next seventeen years, working on various aspects of quantum mechanics and on the general theory of relativity, as well as pursuing studies in Indian philosophy. He also became intensely interested in biology, and in particular the application of quantum mechanics to the foundation of life. His little book, *What Is Life?* persuaded a number of young physicists to turn their interest to biology, where some of them became very productive and prominent (Francis Crick among them). In 1956 Schrödinger returned to Vienna, which welcomed him warmly and showered him with honors. He retired from the university in 1957 and died of heart disease in the Tyrolean village of Alpbach in 1961.

Paul Adrien Maurice Dirac, the youngest of the three creators of quantum mechanics, was born in 1902 in Bristol, England, his mother British and his father an immigrant from French Switzerland. From childhood, he was a loner, preferring to think by himself while taking long walks or gardening. He excelled in science and mathematics at school, but although his pathbreaking later work in physics would be strongly motivated by aesthetics, he tended to neglect literature and the arts. At the age of sixteen Dirac entered Bristol University to be trained as an electrical engineer, while also attending lectures on philosophy, followed by two years of studying mathematics there. In 1923 he transferred to St. John's College at Cambridge, where he became fascinated with the general theory of relativity and for the first time learned about the Bohr-Rutherford atom, earning his Ph.D. degree in 1926.

After spending a year traveling and visiting all the most active research centers in theoretical physics, including Copenhagen and Göttingen, Dirac was elected a fellow of St. John's College and in 1932 appointed the Lucasian professor of mathematics at Cambridge University, a position he retained until 1969. In 1937 he married Margit Wigner, sister of the prominent theoretical physicist Eugene Wigner, whom he had met while visiting the Institute for Advanced Study in Princeton; they had two daughters. In 1971 he moved to the United States, becoming professor of physics at Florida State University. Dirac died in Miami, Florida, in 1984.

Now to the theoretical structure that the three young pioneers erected. Animated by two primary desires, to construct a "mechanics" à la Newton out of mathematical elements but using only directly observable entities, Heisenberg began working with mathematical arrays of Einstein's transition probabilities for Bohr's quantum jumps between various allowed states of atomic systems, arrays that mathematicians call matrices. He soon discovered, to his consternation, that the multiplication of these matrices was not commutative, that is, xy was not necessarily equal to yx, but he forged ahead, and what emerged out of this strange and unintuitive set of rules, with the help of Max Born and a young expert on matrices, Pascual Jordan (1902–1980), was a coherent theory they called matrix mechanics.

One of Heisenberg's first important results emerging from the new theory, after extensive correspondence with Pauli, was based on analysis of experimental measurements. Published in 1927, it came to be known as Heisenberg's uncertainty principle (or, more correctly translated, indeterminacy principle; he regarded it as a fundamental statement of indeterminacy at the submicroscopic level). Suppose an experiment is set up to determine the position of a particle with an accuracy A and, at the same time, its momentum with an accuracy B; then the two error limits A and B cannot both be made arbitrarily small. The best that can ever be achieved is that their product equal Planck's constant h: $AB = h$. In other words, if you want to know the position of the particle extremely accurately, making A very small, then you cannot at the same time demand to know its momentum extremely accurately; the best you can achieve is $B = h/A$.

Heisenberg's conclusion from this surprising principle was that quantum mechanics could not be deterministic: in order to predict the future behavior of a particle, both its initial position and its initial momentum had to be known, but the indeterminacy principle prevented you from knowing both precisely. In fact, even to speak of a particle's motion made little sense, since this would presuppose an exact simultaneous knowledge of both its place in space and its velocity, and such knowledge could not be obtained. As soon as Bohr learned of Heisenberg's new result, which was applicable to any pair of physical variables technically called conjugate, he generalized it, transforming it into a principle of complementarity of wide philosophical sweep but little concrete basis.

As Heisenberg, in Göttingen, was struggling with the unfamiliar mathematics of matrices, he consulted the great mathematician David Hilbert at the same university, who told him that in his experience, matrices usually arose in the context of differential equations. While this remark seemed very puzzling to the young physicist, it turned out to be extraordinarily prescient. Schrödinger was up to exactly that: a formulation of quantum mechanics in the language of differential equations, which turned out to be equivalent to Heisenberg's matrix mechanics.

Meanwhile, a new idea had come to the French physicist Louis-

Victor de Broglie (1892–1987), a doctoral student at the Sorbonne whose reputation in Copenhagen was rather mixed after some run-ins with Bohr. Mulling over the strange notion, put forward by Einstein, that light had a dual nature, manifesting itself both as a particle (photon) and as a wave with interference effects, it occurred to him that this wave-particle duality might be a general phenomenon: perhaps the entities we know as particles, such as electrons, have a wave nature as well. De Broglie postulated that the concomitant wavelength λ should be related to the particle's momentum p by $p = h/\lambda$, where h is Planck's constant. If that were the case, Bohr's postulate that the allowed orbits of an electron in an atom are those for which their angular momentum is an integer-multiple of h would become the plausible assumption that the electron should form a standing wave in its orbit; that is, the length of the orbit should be just a whole-number multiple of the electron's wavelength.

If Copenhagen was cool to de Broglie's notion, Einstein was enthusiastic, and again he was proved right. It was experimentally confirmed in 1927 by the two American physicists Clinton Joseph Davisson (1881–1958) and his assistant Lester Germer (1896–1971), as well as later in the same year by the Scottish physicist George Thomson (1892–1975), son of J. J. Thomson. They all found that a beam of electrons exhibited the same kind of interference phenomena first observed by Thomas Young that had definitively demonstrated the wave nature of light.

Stimulated by de Broglie's idea of particles as waves, Erwin Schrödinger attempted to emulate an elegant formulation of Newton's laws of motion by William Rowan Hamilton that bore much resemblance to the laws of wave optics. The result was that he envisioned the motion of particles governed by a "wave function," for which he devised a differential equation. Although wave mechanics—the version of quantum mechanics centered on the Schrödinger equation published in 1926—appeared to be entirely different from Heisenberg's matrix mechanics, it did not take long for Hilbert's judgment, that matrices usually arise in the context of differential equations, to turn out right. The two were mathematically equivalent; one could be readily translated into the other. However, for purposes of actual calculations and comparison with experimental observations, Schrödinger's version would reveal itself to be very much more useful; it also had the added advantage of being more intuitive and easier to visualize.

But the important question of the *meaning* of the wave function remained. Whereas Schrödinger's initial idea was that it simply indicated particles were somewhat spread out in space rather than being concentrated at a point, this view could not be long maintained: his equation forced the wave function to spread very quickly all over space, and no particle could be expected to do that. It was Max Born, the mentor of both Pauli and Heisenberg, who offered the interpretation that was finally accepted.

Born entered the world in 1882 in Breslau, Germany (now Wroclaw, Poland), where his father was a professor of anatomy at the local university. He received his doctorate in physics and astronomy in Göttingen in 1907 and, after military service during World War I, he was appointed professor of physics at the University of Frankfurt-am-Main. But in 1921 he moved to Göttingen, which he managed to transform, for the next twelve years, into a world center of physics, second only to Berlin (which had Einstein). Because he was Jewish, he left Germany in 1933 for Cambridge, subsequently becoming professor of natural philosophy at the University of Edinburgh in 1936 and a British citizen in 1939. Upon his retirement in 1953 he returned to Göttingen, where he died in 1970.

Born made a number of important contributions to quantum mechanics (a term he coined), but none rivaled his crucial interpretation of the physical meaning of Schrödinger's wave function: its square denoted the probability of a particle's location. (Strictly speaking, the value of the wave function of a particle is a complex number, and the square of its absolute magnitude at a point in space is the probability density of finding the particle there.) Probability thus came to be located right at the heart of quantum mechanics, a characteristic strengthened by Heisenberg's indeterminacy principle, which made statistics into an intrinsic feature of every measurement process.

Dirac entered into the development of quantum mechanics in 1925, at a time when the two quite disparate versions just invented by Heisenberg (with Born and Jordan) and by Schrödinger were making the rounds. He immediately reformulated Heisenberg's matrix mechanics into what later would be called operator language, though he called the operators q-numbers, their main characteristic being that they did not commute (xy was not always equal to yx). Like Schrödinger, he was guided by classical formulations of Newton's laws of motion, but his model was Poisson's rather than Hamilton's way. His transformation theory was a very elegant and general form of quantum mechanics that started from the Newtonian equations of motion, quantized them by replacing all the physical variables by q-numbers, and ended up in equations that encompassed both Schrödinger's and Heisenberg's versions.

By 1927 the trio of Heisenberg, Schrödinger, and Dirac had completed, up to a point, the non-relativistic formalism of quantum mechanics. What had not yet been accomplished was to take the special theory of relativity into account, nor could the equations accommodate the spin of the electron in any way other than ad hoc. Schrödinger had tried his hand at recasting his theory so as to make it conform to relativity, but his new equation led to predictions for the spectrum of hydrogen that disagreed with experimental data. A novel idea was needed, and Dirac provided it.

Many physicists, Schrödinger and Einstein among them, were guided in their search for new insights by aesthetic considerations; beauty was an important criterion for them. This does not mean that they ignored experimental facts, but they were able to generate abstract mathematical ideas relying on their personal sense of beauty and to retain their faith in being right even when, among the always-present welter of initially confusing and unsorted experimental observations, some appeared to prove them wrong. Dirac made no bones about being strongly influenced by aesthetics; as he wrote in capital letters on a blackboard during a lecture in Moscow in 1955, "Physical laws should have mathematical beauty."[4] And the relativistic electron equation he found is universally acknowledged by physicists to be remarkably beautiful. But more important, the predictions based on it turned out to agree extremely well with experimental results.

Fully accommodating the special theory of relativity as well as taking the spin of the electron into account in a completely natural manner, the Dirac equation nevertheless appeared to contain a major flaw: his equation had solutions for positive as well as infinitely many negative energies; there was no ground state in which the electron could remain without jumping further downward, emitting photons. Availing himself of Pauli's exclusion principle, Dirac found an ingenious solution to this problem: he postulated that the vacuum contained an infinite sea of electrons occupying all the negative-energy states, thus preventing any electron of positive energy from descending there. The bottom of the positive-energy states therefore serves as the stable ground state.

This neat picture came with a price, however. Just as it was possible for a photon of the right energy to hit an electron in an atom at any level, thereby being absorbed and kicking the electron to a higher level, so a photon could hit one of the electrons in the negative-energy sea, raising it to a positive energy, while the photon disappeared. The result would be the appearance of a pair consisting of an electron of positive energy and a hole in the sea of negative-energy electrons. This hole, being the absence of a particle of negative energy and negative electric charge, would then act like a particle of positive energy and positive charge. At that time no such particles were known to exist, but it took only a few years for experimenters to catch up with Dirac's beauty-guided ingenious insight into nature.

In 1932 the American physicist Carl David Anderson (1905–1991) discovered the positron, a particle like the electron in every respect, except that its electric charge was positive and of the same magnitude as that of the electron. In later terminology, the positron was the electron's antiparticle, the first such antiparticle to be discovered. Thus, completing its triumph, the Dirac equation contained the prediction of pair creation: under certain conditions, a photon could disappear and give rise to an electron-positron pair, a direct manifestation of Einstein's famous equivalence of mass and energy expressed in the formula $E = mc^2$. This phenomenon would not only soon be observed but would, in more general form, turn out to be of enormous importance in the future development of physics.

With the skeleton of the quantum mechanics of particles now complete, it remained to be fleshed out both in its interpretation (after all, it relied to an inordinate extent on probability) and in its mathematical machinery. We will look first at the major mathematicians who provided the needed structure. At the head of the line was David Hilbert, undoubtedly the most influential mathematician of the twentieth century.

Born in 1862 in Königsberg, East Prussia (now Kaliningrad, Russia), Hilbert inherited his interest in mathematics from his mother. After receiving both his high school and university education locally, he obtained his doctoral degree from the University of Königsberg in 1885, after which he traveled to Paris and Leipzig. He returned to Königsberg to launch his academic career, and became professor of mathematics there in 1893. In 1895 he accepted an appointment as professor at the University of Göttingen, where he remained until his retirement in 1930, though his scientific activity was cut short in 1925 when he fell ill with pernicious anemia, from which he recovered. He died in Göttingen in 1943.

Although politically conservative, Hilbert abhorred nationalism and was one of the few prominent German intellectuals (another was Einstein) to refuse to sign the infamous "manifesto of the ninety-three" during the First World War. Asked by a high Nazi government official after Hitler came to power whether he did not agree that Germany was better off without all those Jewish scientists, he is said to have replied, "Their loss destroyed German science." Hilbert's work in mathematics was enormously wide-ranging, as indicated by his address to the International Congress of Mathematicians held in Paris in 1900, in the course of which he listed 23 unsolved problems, in all fields of mathematics—a list that exerted a powerful influence on twentieth-century mathematics. Many of these problems are still unsolved in the twenty-first century, and every time a question on the list yields to solution, it creates a sensation among mathematicians.

For years Hilbert, both in his own work and in his support of others, actively pursued the aim to place all of mathematics on a firm, formal axiomatic basis, meaning thereby to assure it of permanence. This put him at odds with a strong movement early in the century, called intuitionism, whose most prominent exponent was the Dutch mathematician L. E. J. Brouwer. The intuitionists proposed that mathematics allow only proofs that were constructive. This meant that they denied the validity of a very popular method, in which a theorem was proved by showing that its denial would lead to a contradiction. In Hilbert's publicly stated view, the success of intuitionism would destroy mathematics. The intuitionist movement has now lost momentum, but Hilbert's axiomatic program was dealt a mortal blow from another direction, by the work of the Austrian mathematician Kurt Gödel (1906–1978). Gödel's famous theorem states that in every consistent and sufficiently strong formal axiomatic system there would necessarily arise statements that could be neither proved nor disproved within that system. Restricted to proofs within formal mathematical systems, Gödel's theorem has no relevance outside the foundations of mathematics, the opinion of some prominent physicists notwithstanding.

From the point of view of physics and of quantum mechanics in particular, Hilbert's most important contribution was his work dealing with differential equations, especially his introduction of infinite-dimensional spaces defined by the solutions of such equations. Ever since the refinement of Dirac's formulation of the theory by John von Neumann, the mathematical structure of quantum mechanics, along with the intuition of practitioners emerging from it, has been based fundamentally on Hilbert spaces and the geometrical language associated with them.

Born in Budapest, Hungary, in 1903, the eldest son of a prosperous Jewish banker, John von Neumann was a child prodigy who, at the age of six, could divide eight-digit numbers in his head and exchange jokes with his father in classical Greek. He was privately schooled until he entered the Lutheran Gymnasium, where his mathematical abilities were soon recognized and nourished by individual guidance. After the end of the First World War, he left Hungary (the University of Budapest had a restrictive quota for Jewish students) to study at several German universities, earned a degree in chemical engineering at the Eidgenössische Technische Hochschule in Zurich, and in 1926 received his doctoral degree in mathematics at the University of Budapest. Following post-doctoral work in Göttingen under David Hilbert, he moved to the United States, becoming professor of mathematics at Princeton University and, in 1933, joining the Institute for Advanced Study, also in Princeton, as its youngest professor. He remained there for the rest of his life. Meanwhile von Neumann had married Marietta Kovesi, and they had a daughter, Marina, but the marriage ended in divorce in 1937. The following year he married Klara Dan from Budapest.

A naturalized American citizen, von Neumann began doing consulting work for the U.S. Army, and after the United States entered the Second World War, he joined J. Robert Oppenheimer (1904–

1967) at Los Alamos, New Mexico, in the so-called Manhattan Project to develop a nuclear bomb. The need for large-scale, time-consuming numerical computations at this laboratory awakened his interest in building calculating machines, and immediately after the war he designed and supervised the construction of the first programmable electronic calculator. After the war ended, von Neumann continued to be involved with the design of the hydrogen bomb as well as many other defense projects of the American government. Shortly after President Eisenhower appointed him to membership in the Atomic Energy Commission, he became ill with bone cancer and died in 1957 in Washington, D.C.

The second area, in addition to programmable computing, on which von Neumann left a lasting impression was game theory, which he invented as early as 1926 and for which he wrote, together with the economist Oskar Morgenstern, the very influential treatise *Theory of Games and Economic Behavior*, published in 1944. However, in physics, von Neumann's most important work was his attempt at an axiomatization of quantum mechanics, published in 1932 under the title *Mathematical Foundation of Quantum Mechanics.* While his systematization did not, in the end, prove quite successful, he nevertheless laid the basis for a rigorously mathematical approach to quantum theory, which had evolved from several disparate and partly confused origins.

Beginning with the observation that radioactive emissions spew out particles apparently at random, statistics and probability have become part of the very foundation of physics, initiated by Einstein's transition probabilities for Bohr's quantum jumps. After statistics was introduced into physics in the nineteenth century to explain heat and the behavior of fluids, Boltzmann had shown how these statistics were grounded in an underlying substratum of molecules, each of which followed the Newtonian equations of motion. Disconcerting as such probabilistic laws were to the old guard of physicists, the quantum laws were much more repugnant. Here, there was no assumption of a classical, causal substratum, the ignorance of which necessitated resort to probabilistic laws on the surface. In a quantum worldview, probabilities were at the very heart of the theory.

The now generally accepted physical interpretation of the probabilistic laws postulated for quantum mechanics emerged in 1926 out of long discussions in Copenhagen, primarily between Bohr and Heisenberg, though Dirac and Schrödinger also came for visits and participated, not always in full agreement. The Copenhagen interpretation is based on a renunciation of all assumptions of the reality of entities and processes not observable or measurable: nothing is real until it is measured. If Heisenberg's indeterminacy principle prevents us from precisely measuring them at the same time, simultaneous position and momentum of a particle are meaningless concepts, and hence so is a particle's motion. Its position gains reality when it is measured; its momentum becomes real when it is observed. But to ask "Where is the particle now and how fast is it moving?" makes no sense; such a question is therefore not permitted in quantum mechanics.

The state of a physical system, such as a collection of particles, determined in classical Newtonian mechanics by specifying all their positions and momenta, is determined in quantum mechanics by specifying its wave function. While the behavior of this wave function follows the Schrödinger equation in a deterministic manner—a given state now determines the state at a later time—knowing the wave function does not imply knowing all the physical attributes of the system precisely; for some of them the wave function implies only probabilities. The concept of causality is lost: we cannot trace every event back to an earlier event, or set of events, causing it. What is more, when an observation takes place, the previous wave function is changed—it is "reduced" or it "collapses." This is because after a certain physical property (say, the position) is almost precisely determined, the new wave function must express the fact that others (the momentum, say) have to be correspondingly "smeared out." To put it another way, a previously given probability is "reduced" when a new relevant condition is added.

There is another crucial feature of the new view of the way nature works. The Schrödinger equation has the property—technically speaking, it is linear and homogeneous—that if f and g are two solutions of it, then so is $f + g$; this is called the superposition principle. But on the other hand, the probabilities are the squares of these functions, and $(f + g)^2$ is not the same as $f^2 + g^2$, which seems to contradict the rule for adding the probabilities of independent events if f and g describe independent states.

As might be expected, the ramifications of this way of describing the workings of nature were regarded as objectionable by some physicists, Einstein most prominently among them; other consequences were misinterpreted. At its most intuitive, we have Einstein's famous statement that "God does not play dice with nature." Of course, for believers, God had been primarily responsible for bringing the dice to the table, but Einstein's point was that probabilities could not be fundamental; there had to be a deeper level where causality reigned, so that the probabilities were merely expressions of our ignorance (just like Laplace's view a century and a half earlier).

However, Einstein's objection was more profound than mere unease with probabilities. Although he conceded its success in accounting for experimental data, he did not believe that this theory was capable of completely describing reality, as he pointed out, together with Boris Podolsky (1896–1966) and Nathan Rosen (1909–1995) in a famous paper entitled "Can Quantum Mechanical Description of Physical Reality Be Considered Complete?" published in 1935 in the *Physical Review* (usually called the EPR paper). Using the correlation between the individual states of two particles implied by a two-particle wave function, the article constructed an example in which the position of one particle could be accurately measured and at the same time its momentum could just as accurately be inferred from a measurement performed on the other one, without ever coming near the first and disturbing it. (Heisenberg had intuitively justified his indeterminacy principle as the result of an irreducible disturbance of one of these variables by a measurement of the other.) This demonstrated, the authors claimed, that the simultaneous position and momentum of a particle constituted "elements of physical reality," and yet quantum mechanics did not permit us to define or use them. Hence, they concluded, the theory could not be a "complete description of reality."

Bohr's view, on the other hand, was encapsulated in his statement, "There is no quantum world. There is only an abstract quantum mechanical description."[5] Agonizing over his disagreement with Einstein, whom he admired enormously, he replied with a paper published in the same issue of the *Physical Review.* Its essence was Bohr's contention that "the extent to which an unambiguous meaning can be attributed to such an expression as 'physical reality' . . . must be founded on a direct appeal to experiments and measurements." As these short quotes show, the philosophical issues raised by the new physics were profound.

Most physicists having little patience for philosophy, few paid much attention to the EPR debate. But other questions were equally vexing. The superposition principle appeared to indicate that quantum mechanics violated a basic rule of probability theory and hence of traditional logic. In response, some mathematicians advocated the substitution, in quantum mechanics, of multi-valued logic in place of the traditional two-valued logic, which contained only two "truth values," true and false. However, the apparent conflict between the superposition principle and probability theory was really illusory. As Dirac clearly recognized in his book *The Principles of Quantum Mechanics,* first published in 1930, the superposition principle implied a kind of correlation between otherwise seemingly independent states of a physical system that "cannot be explained in terms of familiar physical concepts," and it is this correlation that gives rise to what would, in its absence, be a contradiction with the addition of probabilities.[6] This characteristically quantum-mechanical correlation is usually called entanglement, and it is one of the counter-intuitive features of the theory. As it seems to reach across space even at long separations, Einstein called it "spooky action at a distance," and its effect is to make quantum mechanics into a non-local theory.

In addition to these perceived difficulties, the various attitudes toward the nature of probability played themselves out among physicists. Some took a strictly subjective, information-based point of view. As Edwin Kemble (1889–1984), one of the first American physicists to absorb the new physics developed in Europe, put it: "If the [wave function] is to be reduced, the interaction must have produced knowledge in the brain of an observer. If the observer forgets the result of his observation, or loses his notebook, the [wave function] is not reduced." The contemporary physicist Bernard d'Espagnat published an article in 1979 in *Scientific American* with the subheading "The doctrine that the world is made up of objects whose existence is independent of human consciousness turns out to be in conflict with quantum mechanics and with facts established by experiment."[7]

While probably most physicists today adhere to an objective view of probability, and hence of the meaning of the wave function, a substantial number are bothered by the frequency interpretation, particularly those who work in cosmology. When dealing with the wave function of the universe, they take the need for an actual ensemble or infinite sequence literally and ask, What does it mean to postulate an ensemble of universes? A relatively recent form of quantum mechanics, based on "histories," is meant to evade this conundrum, but there is no sign that it is catching on.[8] It was precisely to avoid the need for real ensembles that Popper invented his propensity approach to probability theory, but his view does not seem to have caught on much among physicists either.

Stimulated early on by Einstein's skepticism, the question whether quantum mechanics with its irksome probabilities was really fundamental or a mere surface phenomenon, with an unobservable world of hidden variables underneath, has come up repeatedly. John von Neumann thought he had mathematically exorcised this ghost, but his “proof” turned out to be ineffective. However, in 1964 the Irish physicist John S. Bell (1928–1990) elevated the issue to a physically more accessible level by proving a theorem that permitted a clear-cut experimental determination of whether observable correlations between the states of physical systems could be accounted for by orthodox local (without action at a distance) classical means, or whether a quantum mechanical explanation was required. When the needed experiments were performed by A. Aspect and collaborators, quantum mechanics proved to be the winner.[9]

There was, nevertheless, a serious attempt at doing without the quantum, begun by Louis de Broglie and carried to its full flowering by the American physicist David Bohm (1917–1994). Claiming that his theory could reproduce all the results of quantum mechanics, he postulated a world of hidden entities and forces in which causality reigned, so that the probabilities governing observable phenomena were the result of our ignorance of this netherworld. The rub was that the entities populating the hidden realm interacted nonlocally—by an intrinsic action at a distance. This hidden-variable theory evaded Bell’s theorem and Aspect’s experiments by its nonlocality. Could Bohm’s proposed replacement really duplicate with equal accuracy all the successes of quantum mechanics? It has never been fully, seriously tested, simply for lack of interest among physicists. Why should they care about a theory that offered only to be less offensive to our classical intuition, on the questionable assumption that its intrinsic nonlocality was less spooky than the quantum theory, but promised no new results?

Quantum mechanics remains, to this day, the framework theory for our understanding of all physical phenomena. But like Newtonian mechanics, it had to be supplemented with specific laws for the interactions of the entities that it is meant to govern. The discovery of these laws and of more fundamental particles to be governed by them, as well as the extension of the quantum theory to fields, such as the electromagnetic field, will be the subject of the next chapter.

Why are all hydrogen atoms exactly alike? Why does a helium atom weigh four times as much as a hydrogen atom? Why is the spectrum of the yellow light emitted by heated sodium made up of its specific frequencies? One might have expected that quantum mechanics, dealing in probabilities as it does, could lead only to relatively vague or tenuous predictions compared to the definite ones coming out of the classical determinism. Instead, it turned out that the quantum framework, when combined with specific assumptions of forces, was able to yield many predictions of enormously precise numbers, which, when compared to equally precise experimental measurements, proved extremely accurate. While today physicists still harbor various degrees of doubt about some of the ingredients with which to fill the quantum frame, the framework itself, with its probabilities, has shown not the slightest inclination to fail.

eleven

Fields, Nuclei, and Stars

The starting point of any calculation in nonrelativistic quantum mechanics was the assumption of a fixed, given number of particles acted upon by certain given forces. Dirac's relativistic electron equation, however, had demonstrated that when quantum mechanics was combined with the special theory of relativity, this schema no longer worked. As electrons emitted or absorbed radiation, photons were created or destroyed. And since the phenomenon of pair creation made it impossible to keep the total number of electrons fixed, it made no sense to regard the solution of the Dirac equation as a wave function for a single electron à la Schrödinger. The solution of this problem was to invent a quantum theory of fields, a process begun by Dirac in 1927 and continued by Jordan, Heisenberg, and Pauli, as well as by Enrico Fermi. It encompassed two kinds of entities, electromagnetism on one hand and electrons on the other. As Dirac's equation included the forces that electromagnetism exerted on electrons, and Maxwell's equations included electrons as the sources of the electromagnetic field, these two sets of equations would have to be considered together.

The quantization of the electromagnetic field consisted of two steps: the first was to subject Maxwell's equations to the analysis introduced by Fourier, so that one frequency at a time could be dealt with. The effect of this was that the total field was mathematically viewed as consisting of infinitely many harmonic oscillators, each acting like a simple pendulum swinging with its own period. The second step was to treat each of these oscillators quantum mechanically, with the result that their allowed energies made up a ladder of discrete levels. Now, it is the special property of the spectrum of a harmonic oscillator's energies that all its steps from one level to the next are of equal size, call them ΔE. It takes as much energy for the oscillator to jump from its ground state to its first excited state as to jump from the fourth to the fifth. One may therefore view the process in which the energy of the oscillator jumps from one level to the next up as creating a lump of energy ΔE, and when it descends one level as destroying it. Furthermore, the energy lumps are connected with the frequency f of the oscillator by the formula $\Delta E = hf$, just like Planck's. In other words, the quantization of Maxwell's electromagnetic field leads to Einstein's photons—they are simply the quanta of the electromagnetic field.

A similar procedure works for the Dirac equation, the solution of which is now no longer a wave function but a quantum field. This process is called second quantization, as Dirac's original equation is already a quantum treatment of electrons. However, whereas a wave function is a mathematical device for the calculation of probabilities, a quantum field is a condition of physical space. In this case the energies of the quanta of the field automatically have the relativistically correct relation to their momenta; the quanta of this "matter field" are electrons, just as the quanta of the electromagnetic field are photons. The combination of equations for the electron field (including radiation) and for the radiation field (including electrons) is called quantum electrodynamics or QED for short.

A very clever and promising theory it was, which should lead to many observable phenomena. For instance, as the radiation field was capable of producing electron-positron pairs, it could be expected to produce a polarization of the vacuum; that is, the vacuum should have electric properties analogous to those of a piece of matter immersed in an electric or magnetic field. The minimal energy E required for producing a real electron-positron pair equals the relativistic rest energy $2mc^2$ of the two, where m is the mass of an electron and c is the velocity of light, but Heisenberg's indeterminacy principle allows their "virtual" production for short periods of time by photons of lower energy. Similarly, the energy of an electron surrounded by a cloud of such virtual pairs should change, and the energies of atomic levels should shift by small amounts.

The precise magnitudes of all such effects, called radiative corrections, should be calculable by means of the equations of QED. However, as soon as physicists tried to perform these calculations they discovered that all the numbers came out infinite. There were always either too many contributions from "soft" photons (photons of low energy) or too many from "hard" ones (photons of high energy). One of the sources of these infinities was the fact that the ground state of a harmonic oscillator of frequency f has the energy $\frac{1}{2}hf$, and since there are infinitely many of them, they add up to an infinite energy. QED stood at an impasse for more than fifteen years, and physicists were beginning to doubt that the theory would ever be able to produce sensible, verifiable predictions, until it was rescued by two Americans and a Japanese, all working independently and causing great excitement in the physics community in the late 1940s.

Julian Seymour Schwinger was born to Jewish immigrants from eastern Europe in 1918 in New York City. The family was prosperous, as his father's clothing business was doing well. Though precocious, studious, and extremely quick as a child, Julian stood in his older brother's shadow. After attending Townsend Harris, an outstanding high school associated with the College of the City of New York, he entered CCNY, intensely interested in physics, especially in the papers of Paul Dirac, and co-authored his first publication in the *Physi-*

cal Review at the age of seventeen, quickly following it up with several others.

Despite a mediocre academic record at CCNY, Schwinger transferred to Columbia University with the help of Isidore I. Rabi (1898–1988), who recognized his outstanding ability. Graduating from Columbia in 1936, he stayed there as a graduate student, earning his Ph.D. in three years at the age of twenty with a dissertation he had essentially written as an undergraduate. Two years at the University of California at Berkeley with J. Robert Oppenheimer followed, then two years as an instructor and assistant professor at Purdue University, before he joined the Radiation Laboratory at the Massachusetts Institute of Technology to do war-related work on radar and microwaves. His formulations and solutions of wave-guide problems dating from that time turned out to be very influential.

After the end of World War II, Schwinger moved from MIT to Harvard, where he became a full professor in 1947, and married Clarice Carol. A very shy and gentle man with nocturnal work habits, Schwinger attracted many doctoral students, and his lectures were legendary for their elegance and clarity. In 1972 he moved to California, becoming professor of physics at the University of California at Los Angeles; he died of cancer in 1994.

Richard Phillips Feynman was born in 1918 in the borough of Queens of the city of New York, his father a Jewish immigrant from Russia employed as a sales manager of a clothing company, his mother from a well-to-do American family; neither of his parents was college educated. Encouraged early on by his father to be curious about why things are the way they are, as a pupil attending local schools he particularly enjoyed chemistry and mathematics. For his undergraduate studies he went to MIT, initially intending to major in mathematics, but soon switched to physics and impressed his teachers with his completely independent and innovative way of solving problems. He published his first paper in the *Physical Review* and graduated in 1939, then enrolled at Princeton University, where the young American physicist John Archibald Wheeler (b. 1911), who had just joined Princeton's faculty as an assistant professor, took him under his wing, guiding him to his Ph.D. Immediately after receiving his doctorate in 1942, he married Arline Greenbaum, whom he had known since the age of thirteen or fourteen. As he was fully aware, Arline was ill with Hodgkin's disease, and she died three years later.

In 1942 Feynman joined the Manhattan Project at Los Alamos, where the first nuclear bomb was being developed under Oppenheimer's directorship, and where he found himself in the company of Enrico Fermi, Edward Teller (1908–2003), Niels Bohr, Hans Bethe, John von Neumann, and other luminaries in physics and mathematics. After the war, he received many offers and accepted an assistant professorship at Cornell University, where Bethe, whom he greatly admired, taught physics. He soon acquired a wide reputation for a remarkable teaching talent, in an intuitively appealing and flamboyant style, laced with humor and showmanship, instructing and inspiring both graduate students and freshmen. After a year of teaching in Brazil, he moved to California, married Mary Louise Bell, whom he had met at Cornell, and became professor of physics at the California Institute of Technology. His second marriage ended in divorce in 1956. Four years later he married Gweneth Howarth, an English woman he had met in Geneva. This marriage lasted until his death, producing two children, Carl and Michelle. In 1986 Feynman became widely known to the public as a very critical and outspoken member of the commission formed by President Reagan to investigate the explosion of the space shuttle *Challenger.* Already suffering from abdominal cancer at that time, Feynman died in Los Angeles in 1988.

The third member of the trio that saved Dirac's QED and transformed it into the numerically most successful physical theory in history, Sin-itiro Tomonaga, was born in 1906 in Tokyo, the oldest son of a professor of philosophy. Sickly as a child but fascinated by a great variety of chemistry and physics experiments, he attended a renowned school in Kyoto after his father secured a professorship at the Imperial University in that city. Matriculating at Kyoto University, he learned quantum mechanics by reading the papers of Heisenberg, Schrödinger, Jordan, Dirac, and Pauli, obtained the Japanese equivalent of a bachelor's degree in 1929, and remained there as a research student for another three years. Then Tomonaga moved to Tokyo to become assistant to Yoshio Nishina (1890–1951) at his Riken Science Research Institute, where he remained until 1940, save for a couple of years with Heisenberg in Leipzig, working primarily on nuclear physics. He obtained his doctoral degree in 1939 from Tokyo University and also collaborated with Nishina and others in translating Dirac's book on quantum mechanics into Japanese. In 1940 he married Ryoko Sekiguchi, with whom he would have three children. A year later he became professor of physics at Tokyo College of Science and Literature (which was later renamed Tokyo University of Education).

Essentially isolated from the rest of the physics world during the war years, Tomonaga worked on wave guides and microwaves, all the while thinking about QED. When the war ended, he was at the center of the effort to re-establish the theoretical-physics community in the abysmal conditions in Tokyo. Remaining at the Tokyo University of Education and recipient of many honors, he served as president of the university from 1956 to 1962 and became director of the university's Institute of Optical Research and president of the Science Council of Japan from 1963 to 1969, at which time he retired. Tomonaga died in Tokyo in 1979.

Two specific experimental results obtained in 1947 were the impetus for a concerted effort at making QED a viable theory, capable of yielding precise predictions. Both results were measurements of the frequency of the radiation emitted in a transition between atomic energy levels and therefore were capable of high precision, if done with sufficient care and ingenuity. They were performed in full recognition that they might test the existence and exact values of "radiative corrections," provided that theorists could find a way of calculating them by means of QED.

The first, carried out by the German-born American experimental physicist Polycarp Kusch (1911–1993) together with H. M. Foley, allowed an inference on the precise value of the magnetic moment of the electron, that is, the strength of its magnetism. That strength was predicted by the Dirac equation to have a value of 1 Bohr magneton, but QED was expected to change it by a small amount, and the most precise number deduced from several experiments was 1.001146 Bohr magnetons (with an experimental uncertainty of 0.000012). The second, performed by the American physicists Willis Lamb, Jr., (b. 1913) and Robert Retherford (1912–1981), measured, to within a possible experimental error of 0.1 percent, the splitting of two hydrogen levels that, according to the Dirac equation, should have identical energies but which QED was expected to separate, and was named the Lamb shift after the experiment. A theoretical calculation by means of QED had been attempted by Hans Bethe, and although he came close, he had been stymied by the usual infinities. It was Julian Schwinger who, in 1947, found a way of surmounting the difficulties with QED and published theoretical calculations for both the Lamb shift and the magnetic moment of the electron that agreed with their experimental values. The most precise value of the magnetic moment of the electron calculated in 1950 by means of QED was 1.0011454 Bohr magnetons, which agrees with the experimental result to seven significant places.

The method invented by Schwinger and essentially simultaneously by Richard Feynman and Sin-itiro Tomonaga (though in Tomonaga's case without the stimulus of the experiments of Kusch and Lamb, of which he was unaware at the time) came to be known as renormalization. It consisted of sidestepping any attempt at using QED to predict radiative corrections to the mass and the charge of the electron and instead simply employing their experimentally determined values, a procedure made unambiguous by the rigorous demands of the special theory of relativity. When the calculation of these two specific radiative corrections was sidestepped, all other radiative corrections turned out to be free of infinities. The English-born American physicist Freeman Dyson (b. 1923) shortly thereafter showed that this renormalization method of QED could be expected to work in all other future calculations of radiative corrections as well.

These calculations of numerical predictions of QED, though agreeing with experimental data extremely well, were of course not exact. The theory, made nonlinear by combining equations for the quantized electromagnetic field with electrons as sources and Dirac's equation for the electrons subject to electromagnetic forces, was far too complicated for that. Instead, they used perturbation theory, the method originally invented by Laplace and introduced with great effect into quantum mechanics by Schrödinger and others. Its effectiveness here depended primarily on the small value (approximately $1/137$) of the fine-structure constant, usually denoted by the Greek letter α, a dimensionless measure of the strength of the electric charge of the electron. ("Dimensionless" means that it does not depend on the units employed in the measurements of the electronic charge, the velocity of light, and Planck's constant, all of which enter into its definition; α is one of the fundamental constants of the universe.)

Lengthy as all such calculations of radiative corrections are, Feynman's ingenuity led him to introduce an intuitive graphical procedure that proved very helpful and became extremely popular among physicists. Each individual part of a long computation was identified with a Feynman diagram almost as though it described an actual physical process involving electrons and photons. Feynman himself, for whom, just as for Heisenberg, particles rather than fields were the primary objects of nature, did ascribe more physical reality to the processes depicted by his diagrams than did many others, for whom they were nothing but calculational devices (Fig. 14).

Let's go back now to about 1930 in order to catch some experimental discoveries and the inventions that made them possible. One of these was the 1932 discovery by the English physicist James Chadwick (1891–1974) of the neutron, a nuclear constituent whose existence Rutherford had suspected for some time. As these particles are electrically neutral, they were hard to detect directly, and Chadwick inferred their emission in collisions of alpha particles with beryllium atoms from an appropriate amount of missing energy, just as the existence of the neutrino had been inferred by Pauli. (The neutrino, being massless, or nearly so, was, however, even more elusive than the neutron.) With a mass almost equal to that of the proton, though slightly heavier, the neutron finally completed the inventory of the atomic nucleus.

It was now clear that if the number of protons in the nucleus determined an element's atomic number, its position on the periodic

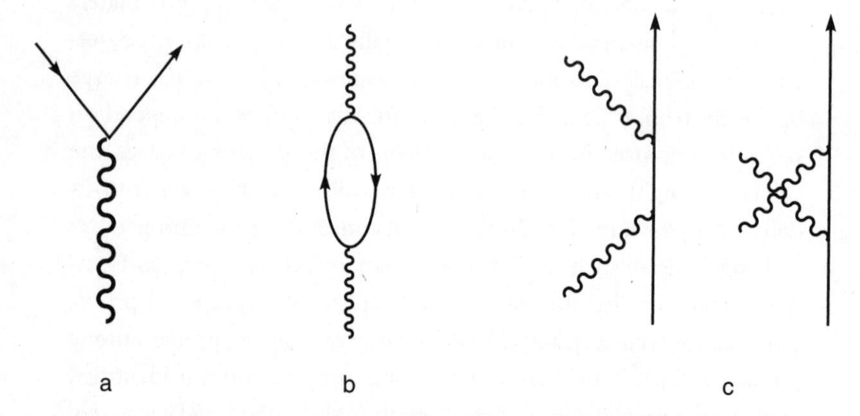

Figure 14 Examples of Feynman diagrams. The wiggly lines are photons, and the solid lines are electrons if the arrow points upward, or positrons, if downward. (a) The creation of an electron-positron pair. (b) A "virtual" pair created and annihilated, a process that contributes to vacuum polarization. (c) Two diagrams for the scattering of an electron by a photon: the Compton effect.

table, and all its chemical properties, the total number of protons and neutrons (they would later both be referred to as nucleons) made up its atomic weight. (Naturally occurring elements, usually made up of mixtures of different isotopes—atoms whose nuclei have the same number of protons but somewhat different numbers of neutrons—have atomic weights that are generally not whole numbers.) Attempts to understand the structure of this nucleus and the process giving rise to radioactivity, which had begun as soon as Rutherford discovered it, would preoccupy a large part of the physics community for the next forty years or so.

As the Rutherford-Marsden-Geiger experiment had indicated, the most promising way to learn the structure of atoms and their constituent nuclei was to bombard them with other particles. But to penetrate into the interior of the positively charged nucleus with other positively charged particles such as protons or alpha particles (since negatively charged electrons were not massive enough to make much of an impact) required that the missiles carry a large momentum so as to overcome electrostatic repulsion. This could be done by accelerating the particles by means of an electric field generated using high-voltage electrodes. In a tour-de-force technical feat, the Englishman John D. Cockcroft (1897–1967) and the Irishman Ernest T. S. Walton (1903–1995) constructed a device to produce sufficiently high voltages for this purpose. Almost immediately, utilizing their Cockcroft-Walton electrostatic generator of 710,000 volts, they succeeded in transforming lithium into helium by accelerating a beam of protons toward a lithium target and detecting the production of alpha particles. From then on, the energies of particles would always be measured in electron volts, eV, that is, the energy a particle with the charge of an electron would have if accelerated by a voltage difference of one volt; soon, the needed unit would be MeV, that is, one million electron volts, and eventually still larger units would be needed.

Another kind of electrostatic generator was invented by the American physicist Robert Jemison van de Graaff (1901–1967). Its first working model, built in 1929, generated 80,000 volts, and his machine was eventually capable of producing 5 million volts. However, electrostatic generators were quickly made obsolete as particle accelerators by the American physicist E. O. Lawrence's (1901–1958) invention of the cyclotron. Lawrence's brilliant idea was to obviate the need for high-voltage generators, bending the path of the particles to be accelerated into a spiral by means of a magnetic field and subjecting them to repeated small electric kicks every time they came around. The crucial fact that made this idea feasible, he realized, was that according to the Maxwell equations, the time it took for a charged particle to circulate once in a given magnetic field remained the same as it went through its accelerating, ever widening, spiraling trajectory. Therefore the electric kicks could be administered at fixed intervals. The first cyclotron he built in 1932 at the University of California in Berkeley had a diameter of only 10 inches but it accelerated protons to an energy of 1 MeV.

Soon, improved versions of the cyclotron were built at universities all over the world, the largest at Berkeley, constructed during the Second World War, with a diameter of 184 inches. That size turned out to be the upper limit, as determined both by the stability of the spiraling beam and by the special theory of relativity applied to the fast-moving protons or electrons. Modifications called synchrocyclotrons and later synchrotrons were invented by the American physicist E. M. McMillan (1907–1991) and the Russian physicist V. I. Veksler (1907–1966); larger and larger models of such machines would produce beams of protons at hundreds of MeV.

Even these energies, however, eventually became insufficient when the purpose of accelerators turned from nuclear physics to the production of particles. A large part of the knowledge gained in physics laboratories in the second half of the twentieth century was generated by scattering experiments, for which accelerators were instrumental. When particles come close to one another or collide, quantum mechanics predicts the probability that they will be found deflected by a given angle, and quantum field theory predicts the probability that a given number of specific new particles will be produced, analogous to pair-creation in the Dirac equation. These probabilities are called scattering cross sections or production cross sections, respectively, and their sizes depend on the forces assumed to be acting and on the applicable conservation laws. Experimental statistical data for scattering or particle production therefore yield information about these forces and the manner in which different fields interact.

Because according to the theory of relativity the energy E required to produce an object of mass m is at least $E = mc^2$, machines with increasingly high energies were required to search for the existence of new, heavier particles and to measure the probabilities of their production by a given process. This is what drove the international race for larger and larger accelerators, most of them constructed for the sole purpose of discovering a specific kind of particle predicted by a promising theory. Their energies would now be measured in units of GeV, billions of electron volts, or even TeV, trillions of electron volts, the diameters of their circular beams in tens of miles, and their costs in hundreds of millions of dollars. Very large linear accelerators, employing microwaves, were also built, primarily to speed up electrons, the biggest one being SLAC at Stanford University, built by the German-born American Wolfgang K. H. Panofsky (b. 1919). Linear accelerators avoided the energy loss through radiation that happened when fast particles moved in circles. The era of big science had begun.

One new idea for improving the power of the impact was to shoot beams of particles at one another rather than direct a beam at a stationary target. This is analogous to a head-on collision in a car, except that the theory of relativity enhances the difference between the two results if their speeds come close to the speed of light. These collisions were achieved by storing the accelerated particles in two ring-shaped tunnels, where they would keep on circulating and eventually hit one another head-on. The Superconducting Super Collider, or SSC, was originally planned to have two colliding proton beams, each of 20 TeV energy, one of its rings with a circumference of 55 miles. This expensive project, to be built in Texas, was finally canceled by the United States Congress. The economic limit of feasible accelerators had clearly been reached, at least for the time being.

A scattering experiment requires not just instruments to speed up particles but also a means to measure the outcome: detectors are needed to find how much the missiles have been deflected or what kinds of particles have been produced, and at what angles. The nature of these detecting devices for charged particles (neutral ones are very hard to detect directly; their presence can usually be inferred only from missing energy and momentum) has evolved over the years, along with accelerators. The first of them goes back to an invention by the Scottish physicist Charles T. R. Wilson (1869–1959), who, fascinated by atmospheric cloud formation, discovered that the essential ingredient for the formation of water droplets (rain) in supersaturated water vapor was the presence of ions rather than dust, as had previously been thought. Applying this idea in 1911, he constructed a clever device for detecting the passing of a charged particle: a transparent chamber, closed with a movable piston and filled with air above a basin of water, so as to contain saturated water vapor. When the piston is suddenly moved outward, lowering the pressure and temperature, the vapor becomes supersaturated, and visible water droplets form at the positions of any ions present. If, at this moment, a fast charged particle passes through, ionizing the air molecules all along its path by collisions, its entire trajectory through the chamber becomes visible as a chain of water droplets. Stereographic photographs of the Wilson cloud chamber reveal the entire track of the passing intruder; moreover, a magnet causes its path to curve, revealing its charge and velocity.

This device was used by the Austrian physicist Victor F. Hess

(1883–1964) in his 1912 discovery of cosmic rays, which eventually revealed themselves as ultra-high-energy charged particles entering the earth's atmosphere from outer space, producing showers of secondaries by colliding with air molecules. The cloud chamber served for many years as an indispensable detector of fast-moving charged particles produced by accelerators. However, in 1952 it was superseded by the bubble chamber, invented by the American physicist Donald Glaser (b. 1926). In this instrument the place of the supersaturated vapor would be taken by a superheated liquid such as ether or liquid hydrogen under pressure, and a chain of tiny bubbles formed by local boiling would show the trajectory of the passing particle. Later, the detector of choice consisted simply of a thick layer of photographic emulsion that would directly retain a record of the passage of charged particles.

Other, more specialized detectors were scintillation counters and Čerenkov counters. The first relied on the fact that zinc sulphide emits a visible flash when hit by an alpha particle, and the second made use of the Čerenkov radiation emitted by any charged object traveling through a fluid at a speed greater than that of light in the same medium. This radiation was experimentally discovered by the Russian physicist Pavel Alekseyevich Čerenkov (1904–1990) and theoretically explained by the Russians Ilya Mikhailovich Frank (1908–1990) and Igor Yevgenyevich Tamm (1895–1971).

All of these detecting devices had the great drawback of requiring the individual scanning of thousands of photographs by technicians trained to recognize the signature of each individual particle on the basis of the thickness and curvature of its track (Fig. 15). As more and more sophisticated and faster computers became available, the previous instruments were eventually replaced by large arrays of spark chambers. Here, the passage of an electric charge between two conducting plates at a high voltage difference produces a spark, recording the arrival of a charged particle at its position and feeding that information directly into a large computer, to be automatically analyzed. Much-superior invisible efficiency thus replaced the more picturesque photographic records.

Nuclear physics was the field that first made the heaviest use of the elaborate instrumentation described. The structure of the outer parts of the atom being pretty well understood, there remained its mysterious nucleus, whose various properties began slowly to emerge. What did it consist of? And how could radioactivity, which Bohr had recognized as originating in the tiny nucleus rather than in the outer layers of the atom, be explained? As to the first question, the nucleus was believed to be made up of protons and electrons until Chadwick discovered the neutron; from then on it was clear that its constituents were protons and neutrons. To account for radioactivity—there was of course no quantum mechanical way to resolve the random

Figure 15 A photograph of the tracks of charged particles in a bubble chamber. A magnetic field makes them curved, revealing their velocity and the sign of their charge.

nature of individual radioactive emissions, which could be predicted only probabilistically—and particularly for the difference between its two forms, alpha and beta, was a lengthy process. The alpha particles emitted by a specific radioactive element all had the same energy, but the energies of the beta particles, strangely, were distributed over a fairly wide range. As Chadwick had discovered in 1914, all that could generally be predicted was the probability of finding an electron of a given energy in a continuous emission spectrum.

Not until 1928 was alpha radioactivity explained on the basis of a typical quantum-mechanical phenomenon. Three American physicists, Ronald Gurney (1909–1953), Edward Condon (1902–1874), and the Russian-born George Gamow (1904–1968), independently recognized that alpha particles would manage to penetrate the force-barrier (the origin of this force was not yet understood) that kept nucleons confined inside the nucleus. Furthermore, the probability of tunneling though this barrier, and therefore their rate of escape, could be calculated by means of the Schrödinger equation. The energy of the alpha particles, on the other hand, was determined by the nuclear quantum levels and thus fixed. The energy distribution of the beta rays, however, remained a continuing puzzle until Pauli's suggestion of the simultaneous ejection of an invisible neutrino: if the energy had to be shared between three particles, an electron, a neutrino, and the recoiling nucleus, the fraction carried away by any one of them depended on their relative directions. The process of beta radioactivity itself, though, remained mysterious until partially illuminated by ideas of Heisenberg and by Enrico Fermi's theory that the guilty party was the neutron, which had the intrinsic property of beta decay.

Born in Rome in 1901, Fermi had a special aptitude for physics and mathematics that was recognized and nurtured from the time he was a teenager. He attended the Scuola Normale Superiore of the University of Pisa, where he obtained his doctorate in 1922. After spending most of 1923 in Göttingen with Max Born in a group that included Heisenberg and Pauli, he went on to Leiden, supported by a Rockefeller fellowship, to study with Paul Ehrenfest and was appointed lecturer in mathematics at the University of Florence in 1924. This is where he made his first major theoretical contribution, the discovery in 1926 of the special statistics governing particles subject to Pauli's exclusion principle. As these replacements of the Maxwell-Boltzmann statistics were discovered independently by Dirac, they have been known ever since as Fermi-Dirac statistics, while the particles to which they apply are called fermions.

The statistics of indistinguishable particles not subject to the Pauli principle, such as photons, had been discovered in 1924 by the Indian physicist Satyendranath N. Bose (1894–1974) and Einstein; they are called Bose-Einstein statistics and the particles obeying them, bosons. Einstein's prediction of the existence of a condensate consisting of many bosons at very low temperature, all congregated on the same quantum level of lowest energy—later called a Bose-Einstein condensate—was not experimentally verified until 1995. Indirectly, on the other hand, the existence of such condensates had already played an important role much earlier in the explanation of the properties of liquid helium. All fermions have a spin angular momentum that is a half-integral multiple of Planck's constant, like the electron, and all bosons have integral spin, like the photon. The later explanation of this spin-statistics connection was one of the major achievements of relativistic quantum field theory. But back to Fermi.

In 1927 Fermi was appointed to the newly established chair of theoretical physics at the University of Rome, and a year later he married Laura Capon, the daughter of an admiral, with whom he would have two children. A highly unusual combination of adroit experimenter, imaginative theorist, and clever problem-solver, in 1927 he invented an influential statistical model of the atom, known as the Thomas-Fermi model after the independent British-born American co-inventor Llewellyn H. Thomas (1903–1992). And in 1934 he made the serendipitous and surprising discovery, which subsequently would turn out to be of great practical as well as theoretical importance, that slow neutrons are much more effective in instigating nuclear reactions than fast ones. In response to the recently enacted racial laws of Fascist Italy—his wife was Jewish—he used the Nobel Prize money he received in 1938 to leave Italy with his family via Stockholm, and emigrated to the United States. (He dissolved the gold medal that came along with the prize money in a jar of acid to hide his valuable possession along the way.)

While Fermi was getting settled at Columbia University, Niels Bohr arrived in the United States from Copenhagen with momentous news. Shooting neutrons at uranium in Berlin, the two German radiochemists Otto Hahn (1879–1968) and Fritz Strassmann (1902–1980) had discovered an unusual reaction. However, because the long-term physicist member of their team, the Jewish Austrian Lise Meitner (1878–1968), had just fled Nazi Germany for Copenhagen, Hahn and Strassmann were unable to interpret what they had seen. Notified of the details, Meitner, who had participated in the experiment until just before its success, and her nephew, the physicist Otto Frisch (1904–1979), together found the correct explanation: the uranium nucleus, struck by a neutron, had fissioned, one of its fragments being the barium Hahn and Strassmann had observed. This discovery sent a tremor through the physics community because, as all knowledgeable physicists realized, it opened the door to a possible chain reaction that could release an enormous amount of energy. The masses of the fragments produced added up to less than the mass of a uranium nucleus, so that the difference would have to be accounted for by kinetic energy and radiation according to Einstein's $E = mc^2$.

Events unfolded following a famous letter from Einstein and the Hungarian-born American physicist Leo Szilard (1898–1964) in 1939 to President Roosevelt, urging governmental support of nuclear research, out of fear that the Germans might develop a uranium weapon first. The pressure escalated after Japan's attack on Pearl

Harbor and America's entry into the war. Having found that the isotope U^{235} was the one fissionable by slow neutrons rather than the more abundant U^{238}, and building on his discovery of the effectiveness of slow neutrons, Fermi led a group of physicists at the University of Chicago in the construction of the first self-sustaining nuclear reactor, or atomic pile. The impact of a slow neutron on a nucleus of U^{235} would break it into lighter fragments, releasing both a large amount of energy and several slow neutrons. These could thereupon in turn break apart other nuclei of U^{235}, thus leading to an explosive snowballing effect unless carefully controlled by neutron-absorbing graphite.

After becoming an American citizen in 1944, Fermi joined the Los Alamos laboratory, where a nuclear chain reaction without such controls was being used for the construction of the first atomic bomb in 1945. Two such bombs, released over Hiroshima and Nagasaki with devastating effects near the end of the Second World War, imprinted the science of nuclear physics on the public imagination for a long time. Upon leaving Los Alamos in 1945 after the war was over, Fermi became a professor of physics at the University of Chicago and continued his work in nuclear physics. He died in 1954 in Chicago. The element of atomic number 100, discovered in 1955, was named fermium, and the unit of length equal to 10^{-15} m is called the fermi.

Shortly after the discovery of the neutron, Fermi developed a theory to explain the nature of beta decay. Modeled after QED, it took advantage of Heisenberg's idea that the neutron and proton were basically the same kind of particle, a nucleon, distinguished only by a two-valued quantum number called isospin—analogous to the spin of the electron but unconnected to angular momentum. (Heisenberg too had proposed a beta-decay theory, but it failed to take the neutrino into account.) The fundamental interaction involved the transformation of a neutron into a proton with the emission of an electron—so that the sum of the electric charges would remain zero—and a neutrino, with an extremely small coupling constant, that is, the analogue of the fine structure constant α in QED. It would later be found that, as befitted such a weak interaction strength, neutrinos could penetrate miles of dense matter without being deflected or absorbed.

The decay of the neutron was actually observed and its half-life measured for the first time in 1950; it is now known to be 615 seconds. The reason why, despite the neutron's instability, not all nuclei containing neutrons are subject to beta decay is that in most of them there is no free state available for the proton resulting from the decay of a neutron; the Pauli principle prevents it from joining another proton in a state already occupied. The receptivity of unoccupied states to emitted protons also determines the different decay probabilities of radioactive nuclei.

While Fermi's schematic idea led to a number of successful predictions over many years, it suffered from a basic theoretical flaw: in contrast to QED, it contained infinities and could not be renormalized. The underlying problem was, essentially, that whereas the basic interaction of QED involved two particles, the electron and the photon, the basic interaction of the Fermi field involved three, the nucleon, the electron, and the neutrino. Some fifty years later, it would be superseded by the electroweak theory.

The picture of the atomic nucleus at this point was one in which, somehow, neutrons and protons were held together quantum mechanically at certain allowed energy levels. These levels were experimentally observable, analogous to atomic levels, by nuclear spectroscopy, except that when an excited nucleus descended to a lower level it emitted a gamma ray, that is, a photon of much greater energy (shorter wavelength) than a photon of light or of X-rays. Generally, the total energy of such a bound system had to be lower than that of its constituents; otherwise there would be enough energy for the system to break apart spontaneously. An alternative would be for a force-wall to keep the nucleons inside. But whereas classically such a wall would be unbreachable, quantum mechanics allowed tunneling through it with a certain probability. This was the explanation of alpha decay proposed by Gamow, Gurney, and Condon.

But the big question remained: what was the nature of the force that held the nucleus together? True enough, Fermi's theory would imply a force between the nucleons, just as QED implied a force between electrons, but that force would be far too weak to overcome the strong electromagnetic repulsion between the equally charged protons. By 1936, proton-proton scattering experiments had shown that, in addition to the Coulomb repulsion, there had to be a strong attractive force between protons, between neutrons and protons, and between neutrons. In the meantime, a field theory had been proposed to explain this force.

The first paper of the Japanese physicist Hideki Yukawa (1907–1981), a friend and classmate of Tomonaga's, published in 1934, contained the proposal of a field theory that would account for the nuclear force, modeled after QED. However, whereas the Coulomb force between electrons had a very long range—it decreased as the inverse square of the distance—the nuclear force between nucleons would have to have a very short range, because outside the nucleus it had never been observed. By the rules of quantum field theory, this implied that the quantum of this field—the analogue of the photon—could not be massless like the photon; in fact, the mass of Yukawa's U-particle could be estimated from the range of the force, which could be no larger than the size of the nucleus. It should weigh several hundred times as much as an electron. The mass of the electron is 0.51 MeV/c^2; so the mass of the U-particle was expected to be between 100 and 200 MeV/c^2. (If energies are measured in MeV, then, according to the Einstein relation, masses are measured in MeV/c^2.)

This was a time of much research on the various components of cosmic rays which, as Victor Hess had discovered, were impinging on the atmosphere from space. In 1936 the two American physicists Carl Anderson and Seth Neddermeyer (1907–1988) had been puzzled by some tracks in their cloud chambers that seemed to belong to particles much heavier than electrons but lighter than protons, and similar ones had been seen by the German physicist Paul Kunze (1897–1986), the French physicist Louis Leprince-Ringuet (1901–2000), and an English team led by P. M. S. Blacket (1897–1974). By 1937 they had been observed by several others too, and Yukawa, as well as Oppenheimer and Robert Serber (1909–1997), speculated that the mesotron seen by Anderson and Neddermeyer might be the U-particle, which was renamed the meson. However, after a few years of uncertainty it became clear that the cosmic ray mesotron could not be Yukawa's meson because it did not strongly interact with nuclei, as was evident from its ability to penetrate through deep layers of earth. It took until 1947 for the real meson to be discovered by the British physicist Donald Perkins (b. 1925) in London, and the team of Cecil F. Powell (1903–1969) and Giuseppe Occhialini (1907–1993), an Italian expatriate, at Bristol. Soon, Yukawa's mesons would be artificially produced at the synchrocyclotron in Berkeley, and their properties were established: they had spin 0, and they came in three varieties, positive, negative, and neutral; the mass of the charged ones was 139.6 MeV/c^2 and that of the neutral one was 135 MeV/c^2; they were renamed pions. The cosmic ray mesotron, on the other hand, with a mass of 105.7 MeV/c^2 and spin ½, like the electron, was renamed the muon.

After the Yukawa force was accepted as the strong attraction that kept nucleons together, there remained the question of the structure of the nucleus. For the purpose of understanding nuclear reactions and fission, George Gamow and Niels Bohr had developed a model based mostly on prequantum physics—further detailed by Frisch and Meitner as well as by John Wheeler—that pictured the nucleus as resembling a liquid drop. Invaded by a slow neutron, it would begin to vibrate and eventually have a certain probability of dividing, its shape changing like a drop of water slowly dripping from a faucet.

A much more detailed quantum-mechanical model was independently devised in 1949 by the German-American physicist Maria Goeppert-Mayer (1906–1972) and the German physicist Hans Jensen (1907–1973). Called the shell model, it resembled Bohr's idea of the electrons in an atom, except that the nucleus lacked the heavy attractive center of the atom. Instead, all the equally heavy nucleons moved along shells (or trajectories) forming the quantum-mechanical energy levels in an average mutual attractive force field. Concurrently, however, Aage Niels Bohr (b. 1922), the son of Niels Bohr, and the American theorists Ben Roy Mottelson (b. 1926) and James Rainwater (1917–1986) developed a quite different collective model, which accounted for certain prominent features of nuclear spectra by viewing the nucleus as a more or less solid body subject to vibratory deformations and rotations. The actual problem of many strongly interacting protons and neutrons forming a heavy nucleus was much too difficult to solve; therefore simplified models yielding well-verified results had to do, even when more than one of them was required simultaneously.

Experimental discoveries during this period included many new isotopes of known elements, most of them radioactive. The first transuranic element, No. 93 (these have atomic numbers above that of uranium, No. 92) was found in 1940 by Edwin Matteson McMillan (1907–1991) and Philip H. Abelson (1913–2004), and they named it neptunium (after the planet Neptune).

The unraveling of the properties of the tiny nucleus at the center of the atom—the fly in the cathedral, as Rutherford had put it—turned out to have wide-ranging applications, not only for making bombs and producing controlled nuclear energy by means of reactors but also in geology and in astrophysics. First of all, as early as 1903 radioactivity was recognized as making a significant contribution to the heating of the earth, thereby removing Lord Kelvin's influential young-earth objection to Darwinian evolution. The first people to point this out were the Irish geologist John Joly (1857–

1933) and Rutherford's former collaborator Frederick Soddy. In 1925 the English geologist Arthur Holmes (1890–1965) used radioactive heating to calculate that the earth had to be at least 1.6 billion years old. Furthermore, as Rutherford had already realized early on, based on the characteristic half-lives of decaying elements, radioactivity could be used to determine the age of geological strata and rocks. This method led to the presently accepted value of 4.55 billion years as the age of the earth, a number Holmes had arrived at by the time of his death.

As far as astrophysical applications of nuclear physics were concerned, the greatest contributor was Hans Bethe, who did important work in nuclear physics in general as well. Born in 1906 in Strassburg (which was part of Germany at that time), little Hans was the only child of Albrecht Bethe, a professor of physiology, and Anna Kuhn, the daughter of a professor. Both were Jews. Precocious in mathematics but frail of health, he was privately tutored until the family moved to Frankfurt in 1915, where his father had been invited to set up a department of physiology at the university; there he attended the Goethe Gymnasium. At the age of eighteen he entered the University of Frankfurt, but two years later moved to the University of Munich, where he earned his doctorate at the age of 22 under the guidance of Arnold Sommerfeld (who nurtured an astonishing number of brilliant young physicists).

Bethe's first appointments were at the University of Frankfurt and at the Technical College at Stuttgart, working with Paul Ewald (1888–1985), but he spent most of the years from 1929 to 1932 visiting Cambridge University and Enrico Fermi in Rome, supported by a Rockefeller Foundation fellowship. Appointed in 1932 as an assistant professor at the University of Tübingen, he was dismissed after one year, as Hitler had come to power; Hans Geiger, with whom he had become friendly in Tübingen, withdrew his friendship at this time. By contrast, Sommerfeld made efforts to find positions for him and other Jewish academics.

After two years in temporary positions in Manchester and Bristol, Bethe immigrated to the United States in 1935, invited as an assistant professor to Cornell University, where he remained for the rest of his life, except for an interruption during the Second World War. In 1939 he married Rose Ewald, his former mentor's daughter, who had emigrated and was a student at Smith College; they had two children, a son, Henry, and a daughter, Monica. Naturalized in 1941, he worked for a time on shock waves with Edward Teller and on the development of radar at MIT. In 1942 he accepted Oppenheimer's invitation to join the Manhattan Project at Los Alamos as director of the theoretical physics division.

In 1946 Bethe left Los Alamos and returned to Cornell to resume teaching and research; he also became active in the disarmament movement, advocating civilian control over nuclear energy and participating in the nuclear test ban discussions in 1958. From 1956 to 1959 he served on the President's Science Advisory Committee. Scientifically productive to the end, Hans Bethe died in 2005 in Ithaca, New York; his last paper was published posthumously.

The two principal problems in astrophysics that involved nuclear physics (and which would turn out to be closely connected) were (1) to account for the origin and the distribution of the elements in the universe and (2) to find the source of the power heating up the stars. In 1936 George Gamow had published a book that included the ideas of thermonuclear reactions in stars and the possible effectiveness of neutron capture in nuclei in building up the heavy elements from helium and hydrogen, which were known to be abundant in stars. A thermonuclear reaction is caused in a high-temperature environment, where the molecules move so energetically that their mutual collisions are able to overcome the electrostatic repulsion between their atomic nuclei to enable them to fuse. In such nuclear fusion, two nuclei combine to form a new, heavier nucleus, which weighs somewhat less than the sum of the weights of the original two. The excess mass is turned into energy according to $E = mc^2$, either in the form of kinetic energy of some expelled protons, neutrons, or other particles, or else in the form of gamma rays.

With temperatures reaching millions of degrees, the interiors of many stars were recognized to have suitable conditions for thermonuclear reactions, and in 1939 Bethe proposed and calculated in detail the probabilities of two specific sequences of reactions. The first, called the pp-cycle, which had already been suggested in a general way by the German physicist Carl Friedrich von Weizsäcker (b. 1912), went through a series of three steps, beginning with the fusion of two hydrogen nuclei and ending with a helium nucleus and two protons, which could then proceed to initiate a new sequence of the same kind. The second, called the CNO-cycle, forms helium out of hydrogen nuclei in a series of six steps that uses a carbon nucleus as a catalyst. Both of these cycles are able to produce enough energy to keep the sun and other stars shining for billions of years. What is more, as a byproduct, the pp-cycle as well as the CNO-cycle gives rise to neutrinos, which are able to escape from the interior of the sun without difficulty; these should be observable on earth and recognizable by their characteristic energy distributions, thereby testifying to the correctness both of Bethe's theory and of our ideas of the constitution of the sun.

A detailed model of the sun's interior was calculated by the American physicist John Bahcall (1935–2005), who then also calculated the precise total number of neutrinos the sun would emit in various energy ranges per second by the CNO-cycle and the pp-cycle. A verification of these numbers would have to wait until methods were found, primarily by the American physicist Raymond Davis, Jr. (b. 1914), to detect the elusive neutrinos. When such procedures were finally developed that could count solar neutrinos in mine shafts underground at a depth to which other particles could not penetrate, their number turned out to be short by a factor of three when compared to Bahcall's model. The solar neutrino puzzle was finally solved in 2003 by means of a new theory according to which there were three kinds of neutrinos, converting into one another in an oscillatory fashion, thereby reducing the number of the kind emitted by the sun arriving on earth. Bethe's proposal, as well as the accepted model of the interior of the sun, were finally found to be right on the mark.

Of course, other stars are also powered by thermonuclear reactions, especially the conversion of hydrogen into helium. The Indian-born American astrophysicist Subrahmanyan Chandrasekhar (1910–1995), usually called Chandra, played a particularly important role in studying the consequences of this phenomenon. The theory he announced in 1935 was this: as a star uses up more and more of its hydrogen and thus cannot emit as much radiation as in its youth, the radiation pressure becomes too weak to overcome the star's own gravitational pull, and it begins to contract and finally collapses into a state of such high density that electrons are unable to move in their normal orbits and atoms cease to exist. A star in this late stage of existence, called a white dwarf, is a very small, dense object of great mass.

However, this fate awaits only relatively small stars like the sun. If its mass is more than 1.4 times the mass of the sun—the Chandrasekhar limit, later more accurately determined to be 1.2 solar masses—the star, unless it can somehow shed some of its weight, will spectacularly explode as a supernova. The remnant of such an explosion is either a white dwarf or else a neutron star, an enormously dense object, two or three times as massive as the sun but with a diameter of only a few miles, consisting of nothing but neutrons. That such entities could exist had been first shown by George Gamow in 1937 and further studied shortly thereafter by the Russian physicist Lev Landau as well as by Oppenheimer and the Russian-born American George Volkoff (1914–2000). Astronomers observed them for the first time in 1967 in the form of radio pulsars, quasi-stellar objects emitting regularly pulsed long-wavelength electromagnetic radiation. (The means of astronomical observation had, since the

1950s, been enormously broadened beyond visible light to include radio waves as well as X-rays and gamma rays; eventually even neutrinos would serve observational purposes.) In 1990 the Hubble Space Telescope was launched, orbiting the earth beyond its atmospheric disturbances and adding greatly to the clarity and reach of what astronomers could observe.

The energy production in stars, involving as it does the "burning" of hydrogen, the lightest element, to make helium, constitutes at the same time the first step up a ladder in which each rung would produce a heavier element either by a process of nuclear fusion or else by neutron capture followed by beta decay. (Adding to a nucleus a neutron, which then decays into a proton while emitting a fugitive electron, raises the element up one step in its atomic number.) Understanding how the heavier elements were built up in the course of the development of the universe took some time to untangle, primarily because there were gaps between some of the atomic masses that were hard to bridge by any known processes. However, the American physicist William Alfred Fowler (1911–1995) and the English astronomer Fred Hoyle (1915–2001), performing elaborate calculations, managed to account for the abundances, from helium to the heavy elements, observed in the stars by means of spectroscopic data. Although calculations of nucleosynthesis, based on quantum mechanics, are probabilistic, they readily lend themselves to reliable statistical predictions for the resulting distribution of elements. Astrophysicists refer to these distributions as abundances. Still, the age-old question remained: was the universe eternal or did it have a beginning?

In 1948 a paper authored by Ralph Alpher (b. 1921), Bethe, and Gamow—Bethe's name was added gratuitously by the playful Gamow to give the list of authors the catchy sound of alpha, beta, gamma—proposed that the universe began with a "hot big bang," consisting of a dense, very high temperature sea of neutrons, from which all the elements would then be synthesized step by step in the course of time.

The same year, Alpher, together with Robert Herman (1914–1997), expanded on this idea but calculated that the early universe would have to have been filled with gamma rays as well, and they came to the remarkable conclusion that as the processes of nucleosynthesis and expansion gradually cooled the cosmos, a remnant of the original gamma radiation should still be visible today in the form of photons with wavelengths distributed like those emitted by a black body at a temperature of about 5° K, putting them in the microwave region.

And indeed, in 1965 the German-born American astrophysicist Arno Allan Penzias (b. 1933) and the American astronomer Robert Woodrow Wilson (b. 1936) serendipitously detected just such an all-pervading background microwave radiation at a temperature of 2.7° K. This discovery laid to rest an alternative cosmology proposed and vigorously advocated by Hoyle. In that theory the universe had no beginning—the name Big Bang for its birth in the alternative theory was Hoyle's derisive coinage—but existed in a steady state in which particles would be constantly created out of nothing to make up for the expansion. Unable to account for the cosmic microwave radiation found by Penzias and Wilson, the steady-state theory was dead.

So by the second half of the twentieth century the universe was understood to have had a beginning some 14 billion years ago in a "big bang," the cooled-down remnant of its radiation dust still detectable in the form of a ubiquitous microwave background and open to examination. However, detailed observations of the distribution of this primordial radiation presented a new problem: it showed a remarkable uniformity that was hard to reconcile with any model of the initial stage of the cosmos. This puzzle was solved in the early 1980s by the American physicist Alan Guth (b. 1947) and the Russian-born American Andrei Linde (b. 1948), who independently suggested that the big bang was followed by a short period during which the newly born universe expanded at an enormous rate—this period is referred to as the time of inflation—wiping out all nonuniformity that may have been initially present. This inflationary scenario implied that the special geometry of the universe, when it settled down, should be flat (exactly on the dividing line between being closed like a ball, with positive curvature, or open with a negative curvature, like a saddle), which more or less agrees with known data.

This was not the end of the story, however. One of the fruits of the new observational power available to astronomers was the discovery in 1998 that some of the very distant supernovae of the kind that can serve as "standard candles" for distance measurements were farther away than expected on the basis of their observed red shift and the Hubble diagram. Unless another explanation could be found—and all such attempts have so far proven unsuccessful—the obvious conclusion was that the expansion of the universe was accelerating. The surprise with which this news was received rivaled the one greeting Hubble's original discovery of the expansion in 1927. Owing to the gravitational attraction of all the mass in the universe, astrophysicists had expected the flight of very distant galaxies to be slowing down rather than speeding up, especially since recent analyses of the motion of galaxies had shown there was much more matter in the universe than what revealed itself by emitting light. The cosmos seemed to be filled with dark matter, the nature of which is still unknown, all of it tending to pull far-away masses to a stop. What could account for their being pushed away faster and faster? One possibility would be Einstein's old cosmological constant, the one he had hastily introduced, to his later regret, in order to keep the universe from expanding. To provide a physical explanation of the otherwise ad hoc cosmological constant, cosmologists are also entertaining the idea of an all-pervading "dark energy" of unknown origin. The issue is still unresolved.

The remaining discoveries of the twentieth century, all their explanations based ultimately on quantum mechanics, deal with the structure of matter and the fundamental constituents of the universe. However, these areas of physics are not to be thought of as entirely separate from those we have been discussing, as was forcefully brought home by a discovery made in 1958. The German physicist Rudolf Ludwig Mössbauer (b. 1929) found that the energies of the gamma rays emitted by nuclei of atoms that are part of a rigid crystal lattice at very low temperature reflect the energies of the nuclear levels more accurately than those emitted by nuclei of freely moving atoms. This is because they suffer a much smaller shift owing to the recoil of the emitting nucleus—in the crystal, the entire heavy lattice recoils, rather than just the individual nucleus, thus making the recoil motion much slower. This Mössbauer effect greatly facilitated the experimental verification of the influence of gravity upon photons that had been predicted by the general theory of relativity—a minute shift, difficult to detect. Notwithstanding the increasing trend toward specialization of scientists, the various areas of physics do hang together and form a whole, frequently complementing one another.

twelve

The Properties of Matter

Prior to the twentieth century, the structure of matter was primarily the domain of mineralogy, and its properties that of thermodynamics. Mineralogists measured the angles between the faces of crystals making up a given mineral and studied all the symmetry types possible in nature (echoes of Platonic solids); thermodynamics dealt with the heat-related behavior of matter. After the discovery of X-rays by Röntgen, the techniques of studying crystalline solids began to change, starting with Max von Laue's discovery that crystals diffracted Röntgen's rays, and continuing with the research of the two Braggs, father and son, who developed the field of X-ray crystallography. That neutrons too could be diffracted by crystals was initially realized in 1936 but could not be fully exploited until the construction of the first nuclear reactor by Fermi in 1942 made substantial neutron beams available. In the meantime, electron diffraction had been discovered by Davisson and Germer in 1927, and that technique was used for the study of crystals as well.

While these experimental investigations added a great deal to our factual knowledge of the structure of solid matter, quantum mechanics began to lead to a theoretical understanding of this structure (particularly when an approximation technique invented by Born and Oppenheimer was employed, based on the large discrepancy between the speeds of the ponderous, slow nuclei and the much lighter, faster electrons in an atom). The first quantum mechanical calculations of the way the atoms in a molecule are held together were done in 1927 by the American physicist Edward Uhler Condon (1902–1974), the Swiss Walter Heitler (b. 1904), and the German—later American—Fritz London (1900–1954). It turned out that the cohesive force arises predominantly from a sharing of electrons among these atoms. (However, there is also an additional interatomic force first discovered in the nineteenth century by the Dutch scientist Johannes Diderik van der Waals [1837–1923] and later explained by quantum mechanics.)

In a comprehensive paper in 1931, followed by the book *The Nature of the Chemical Bond,* published in 1939, the American chemist Linus Pauling (1901–1994) provided a large generalization with a powerful influence on chemistry. Other important contributions to our understanding of molecules were made by the Dutch physical chemist Peter Joseph Willem Debye (1884–1966), whose work on X-ray diffraction transformed such analyses into a much more versatile tool.

The nature of molecules clarified, there remained many bulk properties of matter that begged for explanation, such as their specific heats and particularly the phenomena of magnetization, electrical conduction, and phase transitions—the sudden change of a property as the temperature increases or decreases, such as the freezing and boiling of water.

Owing primarily to the work of Faraday, it had been known since the nineteenth century that the magnetic properties of materials fell into three distinct classes: ferromagnetic, paramagnetic, and diamagnetic. Only three elements possessed the strong—sometimes permanent—property called ferromagnetism at room temperature: iron, cobalt, and nickel. Other materials became very weakly magnetic, to various degrees, when placed in an external magnetic field, in the same direction as the external field if paramagnetic, and opposed to it if diamagnetic. In the late nineteenth and early twentieth centuries, the characteristics of diamagnetism and paramagnetism, particularly their temperature dependence, had been studied by Pierre Curie and Paul Langevin, as well as by the French physicist Pierre Ernst Weiss (1865–1940). Although William Gilbert already knew in the sixteenth century that a permanent magnet loses its magnetism when heated to high temperature, it was Pierre Curie who in 1895 definitively established the existence of a critical temperature (now called the Curie point) above which a given ferromagnetic material becomes paramagnetic through a phase transition. As Weiss found, many paramagnetic substances turned ferromagnetic at very low temperatures; they too had a Curie point. The phenomenon of antiferromagnetism—strong resistance to becoming magnetized—was discovered by the French physicist Louis Néel (1904–2000).

What accounted for the magnetism of all these materials? Ever since Faraday's and Ampère's discovery of the connection between magnetism and electricity and Maxwell's promulgation of his equations, which contained provisions for electric charges as sources of electric fields but none for magnetic poles, magnetic properties of matter were regarded as originating from the motions of charges. This turned out to be correct for diamagnetism, in which the moving charges in a molecule were influenced by an applied magnetic field, but not for paramagnetism, which owed its existence to the magnetic properties of the electrons. At the turn of the century, the Dutch physicist Pieter Zeeman (1865–1943) had discovered that an applied magnetic field could influence atomic spectra. But the resulting shifts of spectral lines, known as the Zeeman effect, remained only partially understood (despite elucidation by Hendrik Lorentz) before quantum mechanics was able to give a complete quantitative explanation, based on the intrinsic magnetism associated with the spin of the atomic electrons rather than on the motion of these electric charges.

Similarly, paramagnetism in matter arises from the permanent spin magnetism of atoms. As the magnetism of the electrons in an atom's inner orbits cancels out in pairs, the spin magnetism of its outermost electron turns an atom with an odd number of electrons into a little permanent magnet of its own. Paramagnetic resonance became a useful new tool: with the atomic magnets lined up by a uniform applied magnetic field, an oscillating magnetic signal at right angles would be strongly absorbed just at the frequency corresponding to the energy needed to yank an atomic magnet around to point the other way. A similar tool, called nuclear magnetic resonance or NMR, was found to exist for the nuclear magnetism, caused by the spin magnetism of the nucleus's constituents in the same way as atoms owe their magnetism to the spin magnetism of their electrons. However, as the nuclear magneton is very much smaller than the Bohr magneton, the nucleus is a much weaker magnet than the electron and contributes little to that of the atom as a whole.

Much of our detailed quantum-mechanical knowledge of the magnetism of matter was developed by the American physicist John Hasbrouck Van Vleck (1899–1980), as well as by Louis Néel. A ferromagnet consists of microscopic crystalline domains, each made up of a large number of atomic magnets, all lining up in the same direction when placed in a magnetic field and remaining that way when the field is removed. However, as the temperature is increased and random thermal motions become more violent, this line-up eventually collapses and the ferromagnetism is lost. What remained unexplained was why ferromagnetism should disappear suddenly, at a critical temperature. Why a phase transition? This was a difficult mathematical problem, made particularly hard to answer because no system consisting of a finite number of elements would exhibit the discontinuous change in its properties as a function of temperature that is characteristic of a phase transition. As indicated by Avogadro's number, the number of molecules in an ordinary piece of matter is so large that real phase transition in physics can be regarded as discontinuous for all practical purposes.

As is often the case when faced with an intractable mathematical problem, physicists make do with a simplified model that contains just enough of the essentials to be relevant but not enough of the complications to be unmanageable. This is what the German physicist Ernst Ising (1900–1998) did with ferromagnetism in 1925, when he set up and treated thermodynamically an infinite schematic array of "magnets," able to point up or down and influencing only their nearest neighbors. Even for this simple model, the proof that at low temperature there was "long-range order"—large domains with all their magnets lined up the same way—which disappeared above a Curie temperature, was a difficult mathematical problem. In one dimension, it was quickly realized, there was no critical point, but as the Norwegian-born American chemist Lars Onsager (1903–1976) eventually proved, for a two-dimensional array a temperature does indeed exist above which the long-range order disappears—an impressive result establishing the existence of a phase transition on mathematical grounds. Alas, it was in two dimensions only; in the physically most relevant case of three dimensions even the Ising model is too difficult, and the existence of a Curie temperature remains an open mathematical question to this day.

The properties of real phase transitions, or critical phenomena, rather than those in models, remained subject to research for many years, and in 1966 the American physicist Leo Kadanoff (b. 1937) found that their graphical description contained a universal feature called scaling, a quantum generalization of what had first been suggested on classical grounds by the Dutch-American physicist George Uhlenbeck. The American Kenneth Geddes Wilson (b. 1936) soon gave this idea a more general mathematical form called the renormalization group. Rooted in quantum field theory, the concept became quite influential in many areas of physics.

The other conspicuous property of matter that begged to be explained was electrical conduction. In 1879 the American physicists Edwin Herbert Hall (1855–1938) and Henry Augustus Rowland

(1848–1901) made the discovery that a magnetic field at right angles to a current-carrying conductor produced a sideways voltage difference orthogonal to both, which came to be known as the Hall effect and allowed the inference that the carriers of electricity were negatively charged. (The electron had not yet been discovered.) Once atoms were understood to make up all substances and were seen to consist of a nucleus surrounded by electrons à la Bohr and Rutherford, it became clear that these carriers of electric currents had to be moving electrons, and the Hall effect was understood as originating from the Lorentz force exerted on the electrons by the magnetic field. However, in 1978, the Polish physicist Klaus von Klitzing (b. 1943) discovered that at very low temperatures the transverse Hall-voltage difference was not proportional to the applied magnetic field but instead rose in discrete steps as the strength of this magnetic field was raised, with wide plateaus of no increase in the voltage. This came to be known as the quantum Hall effect. A partial explanation of this phenomenon was based on quantum-mechanical work of Lev Landau published as early as 1930, but it is still a live research subject.

What remained to be explained was why some materials—notably metals—were conductors, others semiconductors, still others insulators, and why the resistivity of conductors was roughly proportional to the temperature (as had been known for a long time), whereas that of semiconductors decreased as they warmed up. There had also been the surprising discovery of superconductivity.

After the Dutch physicist Heike Kamerlingh Onnes (1853–1926) had in 1908 succeeded in liquefying helium by cooling it down to 4.2° K, he spent the next fifteen years trying to make it freeze, without success. All he found was that at 2.2° K there appeared to be a transition to another kind of fluid he called helium II, but until 1936 no one noticed the remarkable characteristics of this superfluid, as it came to be called: having extremely low viscosity, it can flow through the finest cracks, and it will flow out of any container open at the top by creeping up its wall. In addition, it conducts heat 800 times better than copper. These properties of helium II were discovered and studied in quantitative detail by the Russian physicist Pyotr Leonidovich Kapitza (1894–1984). Their explanation was provided after the end of the Second World War on the basis of Bose-Einstein condensation, by Fritz London, the Hungarian-American Laszlo Tisza (b. 1907), the Russian Nikolai N. Bogolubov (1902–1992), and in greater depth and detail by Lev Davidovich Landau.

Lev Davidovich Landau was born in 1908 in Baku, on the Caspian Sea. He studied at the University of Baku and at the University of Leningrad, where he published his first scientific paper. In 1929, Dau, as he was by then universally called, began an eighteen-month journey abroad, visiting the scientific centers in Europe and meeting all the originators of quantum mechanics except Fermi, the man he admired most. When working with Rutherford in Cambridge, he made the acquaintance of Pyotr Kapitza, but his longest stay was in Copenhagen with Niels Bohr.

After returning to Leningrad in 1931, he moved to Kharkov to head the Theoretical Institute, and in 1937 he married Kora Drobantseva; they had one son, Igor, who eventually became an experimental physicist. This was the time when Landau became interested in low-temperature physics, and promptly in 1937, Kapitza, whom Stalin had summoned back to the Soviet Union from Cambridge, invited him to come to Moscow to head the theoretical division of the Institute for Physical Problems. The following year, however, he was arrested as a German spy, his Jewish background notwithstanding, and spent a year in prison, returning emaciated and sickly. During the Second World War, Landau, appointed professor of physics at Moscow State University, contributed to the Russian war effort, and after the war he participated in the Soviet nuclear weapons program as well as in the development of rockets.

Landau was very unconventional in his personal habits, and his acute critical faculties and sharp tongue were often compared to Pauli's. His influence on Russian physics is impossible to overestimate. Only the exigencies of history and geography prevented the school he created around him from becoming comparable to the earlier one of Arnold Sommerfeld. An automobile accident ended his scientific creativity in 1962; he never fully recovered and died after surgery in 1968.

As for Onnes, he now had liquid helium at his disposal in his laboratory as a cooling medium. In order to study the electrical resistivity of metals at low temperatures, he used this new capability to take them down below 4.2° K and found to his astonishment that at 3° the resistance of mercury for all practical purposes totally disappeared: he had discovered superconductivity, subsequently finding it in lead, tin, and several other metals. An electric current established in a superconducting wire loop would circulate forever without diminution, but a magnetic field of sufficient strength would destroy the superconductivity. What is more, as the two German physicists Walther Meissner (1882–1974) and Robert Ochsenfeld (1901–1993) discovered in 1933, a superconductor immersed in an external magnetic field too weak to destroy its superconducting property will entirely expel the field from its interior—this strange phenomenon is usually referred to as the Meissner effect. Both the zero resistance and the Meissner effect have great potential technological applications that remain to be exploited.

Theoretical understanding of regular electrical conduction, based on quantum mechanics, began with a paper by Arnold Sommerfeld in 1928. Whereas he envisaged the crystals of a metal formed by a regular array of ionized atoms, their outer electrons moving about in the smoothed-out electrostatic atomic field but constrained by the Pauli exclusion principle, a more realistic view was introduced by the Swiss physicist Felix Bloch (1905–1983). His ideas, as well as those of the German-born British physicist Rudolf Ernst Peierls (1907–1995) and the French-born American physicist Léon Brillouin (1889–1969), formed the beginning of what is now called solid-state physics or condensed-matter physics, a field brought to full flowering by the American physicists Frederick Seitz (b. 1911), John C. Slater (1900–1976), and John Bardeen, and the English physicist Sir Nevill Francis Mott (1905–1996), as well as Van Vleck, Wigner, and Landau.

In the resulting picture of a metallic conductor, electrons behave as a Fermi gas—obeying the Pauli principle, they are subject to Fermi-Dirac statistics—in conduction bands above a Fermi surface, below which all quantum states are occupied. The resistance of such conductors is the result of impurities, which introduce into the regularity of the crystals underlying Bloch's theory a certain measure of disorder. The principal developer of the theoretical and mathematical understanding of such disordered materials was the American physicist Philip W. Anderson (b. 1923), who also made important contributions to other areas of condensed-matter physics.

The semiconducting materials, such as silicon and germanium, are situated between conductors and insulators, and their electrical resistivity decreases with rising temperature, rather than increasing like that of metallic conductors, indicating that their conducting properties required a different explanation from those of metals. Their importance resided primarily in the fact that they had been recognized to be suitable for the construction of rectifiers and amplifiers—transistors, as they came to be called—which were much more compact and efficient than vacuum tubes. The development of electronic computers after World War II made such devices particularly urgent. The primary contributors to our understanding of these substances were the German physicist Walter Hans Schottky (1886–1976) and the Americans William Bradford Shockley (1910–1990), Walter Houser Brattain (1902–1987), and John Bardeen.

Bardeen was born in 1908 in Madison, Wisconsin, to Charles Bardeen, a professor of anatomy and dean of the Medical School at the University of Wisconsin, and his wife Althea, an artist and teacher. Intellectually encouraged by his parents, especially in mathematics, which he loved, he skipped three grades and graduated from high school at the age of fifteen, after which he entered the University of Wisconsin, majoring in electrical engineering, with minors in physics and mathematics, graduating in 1928 and obtaining a master's degree a year later. His first job was with the Gulf Research and Development Corporation, thinking up new methods for locating oil deposits. Dissatisfied with this work, he went to Princeton University for graduate studies. He began doctoral research on electrons in metals under the direction of Eugene Wigner but soon interrupted his stay to take up a three-year junior fellowship to Harvard, where he worked with Van Vleck as well as Percy Bridgman (1882–1961), and returned to Princeton to get his Ph.D. in 1936.

Bardeen's first academic appointment was at the University of Minnesota in 1938. For the duration of the war, he was stationed at the Naval Ordnance Laboratory in Washington, D.C., working on the magnetic detection of submarines, and when the war was over, he took a position at Bell Laboratories in Murray Hill, New Jersey, where Brattain and Shockley were pursuing research on semiconductors with the aim of making vacuum tubes obsolete. By 1948 Bardeen's infusion of new ideas into the groundwork laid by the other two led to the first successful construction of a transistor.

By that time, Bardeen had left Bell Labs in 1951 and moved to the University of Illinois as professor of electrical engineering and physics, where he resumed his earlier investigation of superconductivity, laid aside during the war. Although Lev Landau, Vitalii Lazarevich Ginzburg (b. 1916), and Fritz London had taken some steps toward explaining it, this phenomenon still remained puzzling, and Bardeen undertook to unravel it, assisted by his postdoctoral research associate Leon Neil Cooper (b. 1930) and his graduate student J. Robert Schrieffer (b. 1931). Success came in 1956, and its three authors, Bardeen, Cooper, and Schrieffer, were honored in 1972 with the Nobel Prize in physics, the second such award for Bardeen, who remains the only person to have won the prize for physics twice; he had won the first jointly with Brattain and Shockley, for the invention of the transistor. (One of Madame Curie's two Nobels was in chemistry.) Bardeen retired in 1975 and died of heart failure in Boston in 1991.

The BCS theory, as it was soon called, was based on the formation of "Cooper pairs" of electrons, whose mutual attraction results from a distortion of the ambient crystal lattice—somewhat analogous to the polarization of the vacuum in QED. Within a short time it turned out that these Cooper pairs manifested themselves in other characteristically quantum-mechanical ways. As the Welsh physicist Brian D. Josephson (b. 1940) theorized and was quickly found to be correct, they could tunnel through a barrier between two superconductors. The surprising properties of such a Josephson junction turned out to have a number of very useful applications that made quantum mechanical effects directly visible at the macroscopic level.

For the remainder of the twentieth century, many attempts would be made to find materials that became superconducting at higher transition temperatures, the holy grail being a room-temperature superconductor. The principal reason, of course, was practical, as such a substance would have enormously important technological applications. While this goal has remained elusive, the critical temperatures have been pushed increasingly high as time went on, eventually reaching about 150° K. All the high-temperature superconductors are very complicated chemical compounds, and BCS theory does not appear to be adequate for understanding them, but no fully satisfactory substitute has yet been developed.

After this survey of the physics of condensed matter, we turn now to the area of physics dealing with the most basic constituents of the universe, the area that came to be called either particle physics or high-energy physics.

thirteen

The Constituents of the Universe

Before delving into the twentieth-century discoveries about the ultimate constituents of the universe, we have to discuss an area of mathematics that had first become prominent in physics in the context of nuclear physics and whose importance grew as time went on: the field that exploits the consequences of symmetries in nature. (As a symmetry of a system implies that the system remains invariant under certain transformations—rotational symmetry means invariance under rotations—another name for this property is invariance.) We have already touched upon it when we discussed Noether's theorem, as a result of which the conservation laws of energy and momentum in classical mechanics and field theories are consequences of invariance under time shifts and under spatial translations, respectively, and the conservation of angular momentum is a consequence of invariance under rotations.

With the advent of quantum mechanics, such connections between symmetries and conservation laws acquired much more prominence. The strength of classical, deterministic physics was to predict the detailed behavior of systems, such as the sun and the planets. When chance took over, prediction was no longer the primary aim of physics; instead, physicists searched for permanent structures in nature: Why are atoms the way they are? Why is the nucleus the way it is? Why are spectral lines of radiation emitted by atoms and nuclei at those characteristic frequencies? Why do solid materials form crystals? In order to explain the most prominent features of these structures, it was found in many cases to be unnecessary to solve the dynamical equations, such as the Schrödinger equation: they were simple consequences of underlying symmetries. In the language of Aristotle, the second half of the twentieth century replaced efficient causes with formal causes as the dominant explanatory paradigm in physics.

The field of mathematics that turned out to be indispensable for the milking of symmetry properties originated with the young French revolutionary republican Evariste Galois (1811–1832), whose aim was to find methods to solve algebraic equations and to prove under what conditions such solutions were impossible. This was one of the many instances in which a field of mathematics turned out to be extremely useful in physics for purposes other than those for which it had been invented. With the aid of Joseph Liouville, the notes Galois left after his death in a duel at the age of twenty blossomed into what is now called group theory.

A set of mathematical operations is called a group if performing one after another is equivalent to performing a single operation of the same kind, and if each can be undone by an operation called its "inverse." Examples are operations such as rotations, translations, and reflections. In fact, rotations by arbitrary, continuously varying angles form a continuous group; the theory covering such groups was founded by the Norwegian mathematician Sophus Lie (1842–1899) and further developed by the French mathematician Élie Cartan (1869–1951). The physicist who most fruitfully applied group theory to quantum mechanics was the Hungarian-born American Eugene Wigner (1902–1995).

Here is what this mathematical device can do for any quantum-mechanical theory. If all the forces on a given system are invariant under the operations of a certain group, one can, without performing any of the hard equation-solving work required for calculating the numerical values of the various energies allowed for the system, classify them according to the number of states allotted to each energy level. This "multiplicity" of a given energy level equals the dimensionality of the symmetry group's "representation" to which the level belongs, and these numbers can be determined from the properties of the group alone. Furthermore, group theory gives information about transition probabilities from one level to another and indicates when such transitions are forbidden, that is, they have zero probability. Thus, just knowing the symmetries of a system allows physicists to draw valuable conclusions about the structure of its energies and its spectrum, irrespective of the other details of the forces acting in and on it. This ability turned out to be invaluable.

Moving now to physics, recall that at mid-century the elementary particles that had been discovered were the electron (and its antiparticle, the positron), the proton, the neutron, the neutrino, the pion (in its three forms, positive, negative, and neutral), and the muon (negative, and its antiparticle positive). On the basis of the Dirac equation, which was assumed to be applicable, at least approximately, to protons as well as to electrons, everybody expected the existence of an antiproton analogous to the antielectron, even though it could not be found in nature. The first of the new high-energy accelerators, the Bevatron at Berkeley, California, was constructed specifically to speed up protons to an energy at which they could, when colliding with another proton at rest, produce a proton-antiproton pair, and this process was indeed observed in 1955 by the Italian-American Emilio Segrè (1905–1989) and the American Owen Chamberlain (1920–2006). If this did not surprise anyone, its importance lay in the fact that, had the antiproton *not* been found, the implications would have been devastating. The second half of the twentieth century would see an explosion of discoveries of new particles, driven by and in turn driving the construction of ever bigger accelerators.

About the same time as the discovery of the antiproton, evidence accumulated from cosmic-ray showers—and soon from accelerator beams as well, detected by means of cloud chambers and photographic emulsions—that there existed a number of "strange" particles of various masses, electrically positive, neutral, and negative. All unstable, they were regarded as strange because their half-lives were much longer than would have been expected on the basis of their copious production. In other words, if a particle is produced by the same kind of interaction with others that also eventually makes it decay into daughter particles, decay being the same process as production run backwards, a half-life long enough for it to leave an easily visible long track in an emulsion, as these did, should imply that it cannot be easily produced, whereas these objects seemed to be made in relatively large numbers, given enough energy.

The proposed explanation was that there had to be a new conservation law at work in the strong interaction, allowing the easy simultaneous, or "associated," production of two particles with compensating quantum numbers but preventing their decay. That they decayed nonetheless, though slowly, would then presumably be the result of a much weaker interaction—like the one responsible for beta decay—which violated this conservation law.

The particles found were in two classes: the first were hyperons (fermions heavier than protons and neutrons) called Lambda (neutral), Sigma (positive, negative, and neutral), and Xi (negative and neutral); the second, K-mesons or kaons (positive, negative, and two different neutral ones)—bosons lighter than protons but heavier than pions. To account for the observed production and decay regularities, the American physicist Murray Gell-Mann (b. 1929) and the Japanese Kazuhiko Nishijima (b. 1926) devised a scheme of "strangeness quantum numbers," which at the same time implied the existence of two kinds of neutral kaons with different half-lives, a prediction that was soon confirmed by experiments at the Cosmotron accelerator at Brookhaven National Laboratory.

Not all the particles to be discovered in the course of the next twenty-five years, however, lived long enough to leave a visible track in an emulsion. The evidence for most of them was a "resonance" seen in the plot of a scattering cross section. It was one of the results of quantum mechanics that if two particles could form a compound system that remained together for some length of time, the plot of the probability of their scattering when one is shot against the other—their collision cross section—would show a distinct resonance bump, whose width is inversely proportional to the half-life of the unstable compound system: the longer it takes for the unstable system to decay, the sharper the spike in the scattering plot. Therefore the search for new unstable particles concentrated primarily on finding visible resonances in the graphs of cross sections as functions of the energy, ambiguous though their identification often was—especially when the bump in the plot was very broad. And indeed, in a few instances, announced discoveries of particles detected in this manner turned out to be spurious and had to be withdrawn.

What schema was at the bottom of these newly discovered hyperons, and what was the new conservation law that explained their long lives? As we have seen before, a conservation principle was always the result of a symmetry in the underlying equations, and the multiplicities of equal-energy quantum states—there were a total of eight heavy fermions called baryons, comprised of the proton, the neutron, and the six hyperons—could be calculated algebraically by means of group theory. The symmetry group that did the trick, found by Gell-Mann and independently by the Israeli physicist Yuval Ne'eman (1925–2006), was called SU(3), and it had indeed an eight-dimensional "representation" (Gell-Mann called it the "eightfold way") that exactly accommodated the eight baryons. Moreover, the same symmetry group also accounted for the mesons, that is, the three pions and the four kaons, plus a subsequently discovered meson to be called eta. If the underlying field equations were exactly invariant under the SU(3) symmetry, the eight baryons should have the same mass, and so should the eight mesons, contrary to experimental facts. So Gell-Mann and the Japanese-born American physicist Susumu Okubo (b. 1930) devised a way in which that exact symmetry was slightly broken, and they calculated the resulting mass changes, obtaining reasonable agreement with the data. But that was not all.

In the meantime, nine particles had been found which seemed to be excited states of the baryons: one excited state of each of the three Sigmas and the two Xis, and four of the two nucleons, called delta, neutral, negative, positive, and doubly positive. These nine states fit perfectly into a ten-dimensional representation of SU(3), with all the charges exactly accounted for. However, the tenth place of the scheme was experimentally unoccupied, reserved for an unknown negatively charged particle called the omega-minus, with all its quantum numbers and its mass (based on the previously employed Gell-Mann–Okubo mass formula) determined by the schema proposed by Gell-Mann and Ne'eman. Two years after its prediction by Gell-Mann, it was found in 1964 in a bubble-chamber photograph at Brookhaven National Laboratory.

If these discoveries illustrated symmetry laws in nature and their effect on the classification of newly found elementary particles, at about the same time a spectacular discovery violated one of nature's seemingly sacred symmetry principles: invariance under mirror reflection. All the known equations expressing physical laws remained unchanged when a given system was replaced by its mirror image: this invariance led in quantum mechanics to parity conservation, parity being a positive or negative quantum number assigned to a given state. Two of the strange mesons discovered in the 1950s, called theta and tau, presented a puzzle: although they had exactly—to within experimental error—equal masses and half-lives, their intrinsic parities seemed to be different: the theta decayed into two pions and the tau into three. Since the intrinsic parity of the pion was negative and parity was assumed to be conserved, this meant that the theta had positive parity, the tau negative, and—their equal masses and lifetimes notwithstanding—the two could not be the same particle.

The Gordian knot of the tau-theta puzzle was ingeniously cut in 1956 by the two Chinese-born American physicists Chen Ning Yang (b. 1922) and Tsung-Dao Lee (b. 1926), who proposed that tau and theta were the same particle but that their decay violated the law of parity conservation. To bolster their argument that in this weak decay, which they assumed to be caused by the same interaction as beta decay (this was the crux of the matter), parity was not conserved, they proposed that conservation of parity ought to be checked experimentally in other instances of beta decay, and they pointed out several such possible tests. Because parity conservation had always been taken for granted, such specific studies of radioactivity had never been done before. However, later re-examination of older experimental data revealed indications of its violation, which had been ignored as obviously erroneous. When the Chinese-born American physicist Chien-Shiung Wu (1912–1997) quickly performed an appropriate experiment on carbon-12 (the carbon isotope of atomic weight 12), she confirmed what Lee and Yang had suggested: the beta-decay interaction, in fact, strongly violated the conservation law of parity. (Pauli's first reaction to the proposal of Lee and Yang had been, "I can't believe God is a weak left-hander." He soon ate crow.)

Nature's violation of mirror symmetry, which had been thought sacrosanct, opened a Pandora's box. In fact, the emerging weak-interaction theory violated not only P, that is, parity, but also C (a symmetry transformation called charge conjugation, which turns particles into their antiparticles) in such a way that their combination CP stayed inviolate. There remained a third fundamental transformation called T (time reversal). Nature's invariance under T assured that a video tape of a fundamental process run backwards showed an equally possible process, and it was one of the proud results of relativistic quantum field theory that the combination CPT had to remain invariant, even if a particular theory allowed violations of C, P, or T. Theoretical studies had shown that the weak decays of the two neutral kaons were the most promising experimental ways of testing whether CP was in fact conserved, and in 1964 an American team led by James Watson Cronin (b. 1931) and Val Logsdon Fitch (b. 1923) found its violation after a detailed analysis of their data. Presumably this implied that T symmetry also had to go, and indeed their data were subsequently found to imply—without assuming CPT invariance—that time-reversal symmetry was violated as well.

The discovery of strong parity non-conservation in beta decay had a shattering impact on the theory of the neutrino and its interaction with the neutron and the electron, the source of beta-radioactivity. The easiest way to account for the parity violation observed was to return to an older mathematical description of neutrinos with roots going back to the German-born American mathematician Hermann Weyl (1885–1955): the neutrino is an intrinsically "left-handed" particle; its spin rotation together with its direction of motion gives it the screw-sense, called helicity, of a left-handed screw—a property it could permanently retain only if its mass was exactly zero. The mirror image of a left-handed screw is right-handed; this lack of mirror symmetry is why the Weyl theory had previously been discarded. Experimental confirmation of the neutrino's helicity was not long in coming. An ingenious experiment by the American team of Maurice Goldhaber (Austrian-born in 1911), Lee Grodzins (b. 1926), and Andrew William Sunyar (b. 1920) confirmed it in 1958. However, neutrino oscillations, which solved the solar neutrino puzzle, are possible only if the mass of the neutrino is not exactly zero, so helicity cannot be a neutrino's permanent state.

Within two years, Feynman and Gell-Mann reformulated the manner in which this altered neutrino interacted, based on a universal kind of current-mediated force subsequently exploited more fully by Gell-Mann. The resulting theory of beta radioactivity was called the V-A law. It was also independently put forward by the American physicists E. C. G. (George) Sudarshan (born in India in 1931) and Robert E. Marshak (1916–1992). Eventually, this approach was translated into the replacement of Fermi's beta-decay theory by a more conventional field theory, in which a new boson, later named W, acted as an intermediary. There was one hitch: the assumption of its universal nature led to the prediction of the decay of the muon into an electron, with the emission of a gamma ray, at a rate sufficient to have been observed; but that decay had never been seen. This could be explained only if the neutrino involved in the weak muon interaction was different from the one participating in the weak electron interactions. In 1962 the two-neutrino hypothesis was indeed experimentally confirmed at the Brookhaven National Lab and shortly thereafter at the CERN laboratory in Switzerland. The search for the W and an associated particle called Z, however, proved frustrating until they were finally found in 1983 at CERN by the Italian physicist Carlo Rubbia (b. 1934) and the Dutch physicist Simon van der Meer (b. 1925).

The idea of the W boson allowed the use of a field theory analogous to electromagnetism to describe the way leptons—electrons, muons, and neutrinos—interacted with one another, with the W playing the part of the photon. Some ten years earlier C. N. Yang, together with the American physicist Robert Laurence Mills (b. 1927), had shown that the form of the beta-decay force could be explained by a general symmetry analogous to a concept in electromagnetism, long familiar to physicists, called gauge invariance. Field theories embodying this symmetry would henceforth be called Yang-Mills theories, and the idea proved so stimulating that all subsequently invented field theories would follow it.

In the particular instance of the weak interaction mediated by the W boson, its similarity to electromagnetism had led Julian

Schwinger early on to the thought that the weak and electromagnetic actions might be different aspects of the same phenomenon. Whereas initially misleading experimental observations on the W appeared unfavorable, Schwinger's student Sheldon Lee Glashow (b. 1932) carried the idea to fruition, except for one remaining flaw: the gauge-invariance of the theory, crucial for freeing it of infinities like QED, would imply that the W and Z were massless like the photon. This obstacle, however, was finally overcome using ideas partly imported from the BCS theory of superconductivity by Philip Anderson and the Japanese-born American physicist Yoichiro Nambu (b. 1921), as well as those of the English physicists Jeffrey Goldstone (b. 1933) and Peter Higgs (b. 1929), based on concepts called dynamical symmetry breaking and spontaneous symmetry breaking. The basic notion here is that although the equations of a theory may be invariant under a certain transformation, some of their solutions may not be. A classical instance: the Newtonian equations of motion of the planets around the sun are rotationally symmetric, but the elliptical orbits of the planets are not. At the same time, the spontaneous symmetry breaking would generate a new particle, widely referred to as the Higgs particle. In a sense, the theory holds the Higgs particle responsible for the masses of the W and the Z. The principle of spontaneous symmetry breaking would turn out to be a very influential idea; all-pervading as the exploitation of symmetries had become, assuming their spontaneous violation would also turn out to be very handy at some crucial junctures.

Independently of Glashow, the American physicist Steven Weinberg (b. 1933) and the Pakistani Abdus Salam (1926–1996) exploited the Higgs mechanism for the same purpose, unifying the weak and electromagnetic interactions. All these ideas, however, were widely ignored—particularly since they implied the existence of a weak neutral current, which had never been seen in any laboratory—until the Dutch theorist Gerardus 't Hooft (b. 1946) proved that the theories were renormalizable like QED, that is, their predictions were not beset by infinities. Thus was born the electroweak theory. What is more, the missing neutral currents were finally discovered, some of the evidence for them even hiding in misinterpreted old data.

Meanwhile, the strong interactions among particles—those responsible for the stability of atomic nuclei, absent their beta decay caused by the weak interaction—remained to be understood. Although the dynamics involved were still obscure, Gell-Mann's successful exploitation of SU(3) for classifying the strange particles led in the right direction. In 1964 he proposed to exploit the fundamental representation of SU(3), which was three-dimensional—SU(3) is a transformation in three dimensions, though not in physical space—and postulated that all the particles previously regarded as elementary could be made up of three more basic fermions he called quarks (from the line "three quarks for Muster Mark" in James Joyce's *Finnegans Wake*). Each quark would have an electric charge equal to a fraction of the electron's: two of them positive with 2/3 of the electronic charge, and one of them negative, 1/3 its strength. A similar scheme was proposed independently by the Russian-born American physicist George Zweig (b. 1937), who named the particles "aces"; however, "quarks" stuck.

Since no such particles with fractional electronic charges had ever been seen, Gell-Mann initially regarded his picture, according to which each baryon consisted of three quarks and each meson (the pions, kaons, and some newly discovered ones, all with integral spin quantum numbers) of a quark and an antiquark, as no more than a mathematical scheme devoid of reality. Real or not, it also helped to explain why only the eight-dimensional and the ten-dimensional representations of SU(3) were realized in nature. Experimentalists had been searching in vain for evidence of particles belonging to other representations of the same group. However, as independent evidence from scattering experiments, analyzed by Feynman and others, began to indicate that nucleons may be made up of point-like partons—analogously to the way the Geiger-Marsden experiment showed Rutherford that the atom had a nucleus—the notion of quarks acquired reality, though the search for them outside the confines of baryons or mesons has been in vain to this day.

As fundamental building blocks, quarks brought explanatory order to the bewildering, apparently chaotic zoo of new elementary particles that were being discovered at the big accelerators in the form of resonances, just as the discovery of the constituents of atoms had explained the periodic table of the elements. Why quarks would never be found isolated was subsequently argued, though not proved, to be the result of a feature called asymptotic freedom, elucidated by three American physicists, David Gross (b. 1941), David Politzer (b. 1949), and Frank Anthony Wilczek (b. 1951). The attractive force between quarks is assumed to diminish at small distances but to remain constant at large distances—so the intuitive argument goes—as a result of which any attempt to separate them requires enough energy to produce quark-antiquark pairs—thus, mesons—instead of achieving the aim of separation.

A conundrum about the appropriate statistics for quarks nevertheless remained. They needed to have the same half-integral spin as electrons and protons, and therefore should be fermions obeying the Pauli principle, but that did not seem to fit the data. The puzzle was solved by Nambu together with the Korean-born American Moo-Young Han (b. 1934), who suggested the existence of an additional quantum number: each quark came in three different "colors" (nothing to do with actual color; the name is entirely metaphorical, though the concept is precise). Just as the spin quantum number allowed every state in an atom to be occupied by two electrons, one with spin up and another with spin down, instead of by one only, thereby leading to the systematics of the periodic table that agreed well with the data of chemistry, so the color quantum number allowed quarks of different color to occupy the same state, which brought agreement with the experimental particle data.

Also waiting to be explained were intriguing parallels between the hadrons (strongly interacting particles) and the leptons. For one thing, there appeared to be connections between three different kinds of decays, the decays of hadrons with and without changes in strangeness, and the decays involving leptons; this issue was clarified to great effect by the Italian theorist Nicola Cabibbo (b. 1935). For another, there was a striking parallelism between the ways the hadrons and the leptons were classified (a parallelism that was in fact required by the theory in order for it to be renormalizable, that is, free of infinities), if it were not for one missing quark: there were four leptons (the electron, the muon, and their separate neutrinos) but only three quarks. To fill up this hole, Glashow and the American James D. Bjorken (b. 1934) proposed the existence of an additional quark that differed from the others by being "charmed." Increasing even further the complications of particle physics, in 1975 the American physicist Martin Perl (b. 1927) discovered an additional lepton, named tau, whose mass was more than twice that of the proton. Since the tau, like the electron and the muon, was also found to have its own neutrino, this raised the total number of leptons to six. The number of different quarks thus finally also had to grow to six, their "flavors" named Up, Down, Strange, Charmed, Top, and Bottom. (The last two are sometimes also called Truth and Beauty; long gone are the days when new scientific terms were coined in stodgy Latin.) Only the first two are stable; the other four are much heavier and have finite lifetimes.

The experimental verification of the existence of the additional quarks turned out to be difficult and full of confusion. The Top quark was not found until 1995, at Fermilab, and its mass turned out to be approximately 180 times that of the proton (almost as heavy as a tungsten atom). The first Charmed meson, called D, was found in 1976 by the German-born American physicist Gerson Goldhaber (b. 1924). It came as a doublet (hence the D): one neutral, consisting of a Charmed quark and an anti-Up quark, and the other positive, made up of a Charmed quark and an anti-Down quark. Meanwhile, the first evidence for "charmonium"—a composite of a Charmed quark and a Charmed antiquark—was found in 1974 simultaneously by the Americans Samuel C. C. Ting (b. 1936) at Brookhaven National Lab and Burton Richter (b. 1931) at the SLAC accelerator laboratory at Stanford University. As Ting had named the particle he discovered J and Richter had dubbed the one he found Psi, it came to be known as J/Psi. By now the production of Charmed particles by accelerators has become routine, and the quark scheme appears to be complete, though additional flavors cannot be completely ruled out.

Confusing as the road to an understanding of the enormous mélange of elementary particles was, the systematics and classification introduced by the notion of constituent quarks in three colors and six flavors has clarified the picture as much as the Bohr-Rutherford atom clarified chemistry by furnishing a basis for the periodic table.

The existence of quarks as basic constituents of all hadrons, of course, required a quantum field theory to explain the strong forces holding them together, and this theory came to be called quantum chromodynamics or QCD, a name that echoed QED, after which it was modeled. A gauge theory à la Yang-Mills, built on the underlying symmetry of SU(3), it uses the three colors (hence the chromo prefix) of particles in place of the one electric charge employed by electrodynamics, and the quanta of its force-field, the analogues of photons, are called gluons. There are some fundamental differences between QED and QCD, however. The eight gluons, which belong to an eight-dimensional representation of SU(3), just like Gell-Mann's original eight-fold way leading to quarks, themselves carry color and hence directly interact with one another and are able to change the color of a quark. In contrast to photons, they so strongly attract one another at low energy that they can form "glueballs." (No glueball has been experimentally found as yet, however.) At high energy, on the other hand, the strength of the interaction mediated by gluons decreases: this is asymptotic freedom, thought to be responsible for the fact that neither quarks nor gluons can ever be seen roaming freely.

While QCD has succeeded in predicting (mostly post hoc, to be sure) the general structure of the observed hadrons and many qualitative aspects of their production and scattering probabilities, even approximately yielding the hadron masses once the quark masses are given, this theory cannot rival QED in the precision of its calculable predictions of observable results. The principal reason for this failing is that, in contrast to QED, it does not contain a parameter of small numerical value—like the fine-structure constant—to facilitate reliable approximate computations. Nevertheless, the combination of the two quantum field theories, QCD and the electroweak theory (incorporating QED), is now called the standard model of the elementary particles. This model still contains many adjustable parameters, such as the quark masses and the greatly different strengths of the three contributing theories.

The general idea now is that at extremely high energies (or at extremely small distances) the three interactions become equally powerful, and the emerging Grand Unified Theory (GUT, as it is called) is generated by a symmetry that puts quarks and leptons on an equal footing, forcing the quark masses to vanish. These masses would be resurrected by the Higgs mechanism, which would manifest itself in the appearance of a heavy Higgs particle. The overarching invariance would include supersymmetry, a symmetry not previously envisioned between bosons and fermions, uniting particles of integral spin and half-integral spin by postulating that every particle of given spin S has a partner of approximately the same mass whose spin differs from S by half a unit. At this point, no experimental evidence exists for any such partners, and the search for the Higgs particle has so far remained unsuccessful. On the other hand, the solution of the solar-neutrino problem provided evidence that neutrinos could not be entirely massless. Subject to the effects of the Higgs mechanism, neutrinos too would end up with non-zero masses in the standard model.

A further consequence of the envisioned grand unification would be to make the proton unstable, in violation of one of the most basic laws of particle physics: baryon conservation. The Russian physicist Andre Sakharov (1921–1989) speculated in 1967 that the puzzling large preponderance of baryons over antibaryons that has been observed in the universe could be explained by a combined breaking of both CP conservation and baryon conservation at a period in the universe's history when it was out of thermal equilibrium. However, extremely sensitive searches for proton decay have been able to push the half-life of the proton—if it is indeed unstable—beyond 10^{33} years, far beyond the age of the universe, a result that rules out some versions of GUT but not others.

The attempts at grand unification so far mentioned leave out the weakest interaction of all, but the one most important for the structure of the universe as a whole: gravitation. The principal reason for this is that Einstein's general theory of relativity has never been successfully combined with his other brain child, the quantum. This lack of compatibility is particularly troubling because, while in the realm of ordinary particle physics gravity is indeed negligibly small, there is an energy, which Max Planck already recognized as a "natural" unit (it can be calculated just from Planck's constant, the speed of light, and the gravitational constant that enters into Newton's law of gravity) beyond which the strength of the force of gravity rivals the others. This Planck energy has the enormous value of 10^{19} GeV, but it is not very much higher than the energy at which the strong and the electroweak forces are assumed to become comparable and the grand-unifying symmetry is supposed to swing into action. Thus, while at this ghostly energy all the interactions might be imagined united, gravity was still a force apart.

In order to remedy this glaring deficiency in our understanding of nature at the Planck scale of energy (the estrangement between the standard model and gravity), an entirely new class of theories has been pursued, called string theories or, when combined with supersymmetry, superstring theories. First recognized by the American physicist John Henry Schwarz (b. 1941) and the British physicist Michael Boris Green (b. 1946) as naturally incorporating a graviton-like entity (graviton would be the name of the quantum of gravity if ever the gravitational field could be quantized) and therefore promising to lead to a theory of quantum gravity, this research program became extremely active during the last quarter of the twentieth century. Its two most prominent contributors were the American Edward Witten (b. 1951) and the Israeli-born American Nathan Seiberg (b. 1956). This theory replaces point particles with one-dimensional, stringlike objects of Planck length—again a "natural unit" whose length is about 10^{-35} m, far smaller even than nuclear sizes—vibrating in nine space dimensions (plus one time dimension, though the total number of dimensions varies somewhat among some versions of these theories). To account for the fact that the known physical space has three dimensions, the space in which the strings live is imagined as curled up, "compactifying" the extra dimensions and rendering them unobservable.

While the promise of combining gravity with quantum mechanics makes string theories attractive, their drawback is that they cannot be subjected to experimental tests at feasible energies. There is no hope of ever constructing an accelerator reaching 10^{19} GeV, an energy far beyond even those of cosmic-ray particles. As a consequence, string theories are judged primarily by aesthetic criteria. They are admired for their mathematical beauty and elegance, with the ultimate hope that the final "theory of everything" might emerge as both the most beautiful and the only one logically possible. If realized, the definitive answer to Einstein's question "Did God have any choice in the way He constructed the universe?" would be no. Absent such uniqueness and experimental evidence, elegance may have to do.

To account for specific dimensionless numerical fundamental constants contained in the physical laws governing the universe, such as the fine structure constant, other coupling strengths, and the ratios of masses of elementary particles, some physicists have proposed an argument known as the anthropic principle. Its basic idea is that important characteristics of our world, such as the existence of stars and galaxies as well as the stability of many elements, including carbon, are sensitive to the values of these constants. If they were altered even slightly, the universe would be so different that intelligent life could not exist in it. The fact that we humans are here, the argument goes on, explains why these constants have the values they have. This causal reasoning, with its quasi-teleological flavor, can be interpreted in a variety of ways, ranging from the religious to the probabilistic: there may be many universes, all with different values of the fundamental constants, but only one that is inhabited by intelligent beings. The anthropic principle is extremely controversial among physicists but does have some prominent adherents.

Though Aristotle's specific laws were abrogated some four hundred years ago by Galileo and Newton, his search for laws of motion remained the prime motivating force of most physicists for almost two and a half millennia. Today, however, we seem to have reached an era that would be more palatable to Plato. Whether any of the imaginative concepts proposed in recent years will survive in physics (as opposed to mathematics, where string theory has turned out to be quite fertile) remains an open question. One thing is sure: the end of physics is nowhere in sight.

Epilogue

Covering a time span of some six millennia, this book has followed the development of the part of science that was called physiology by the ancient Greeks, natural philosophy at a later time, and now physics. We have seen that the attention of early physical scientists was primarily focused on describing the motions of the heavens and on postulating the constitution of matter. The Greek philosophers began to search for the rules by which objects moved the way they did, and Aristotle was the first to lay down specific causal laws governing these motions, which gradually led to a view of the heavens as moving like clockwork (unless God intervened). However, it took until the first scientific revolution, wrought by Galileo Galilei and Isaac Newton more than two thousand years later, for this view to be unambiguously brought down to earth and applied not only to the heavens but to the falling apple as well.

Aristotle's general search for the rules by which things moved remained central to physics, but his specific laws of motion were found wanting and replaced by Newton in a form that proved implementable by calculations. Newton not only cast his own laws in the form of equations, but he also devised the essential mathematical tools required to solve these equations. By the time of Laplace, nature viewed as clockwork was seen as all-pervasive and could even be imagined, given sufficient data and means of calculation, to be exploited for practical purposes.

The nineteenth century foreshadowed the second scientific revolution by destroying the Aristotelian dream and disconcertingly introducing probabilities into physics, the temple of certainty. A hitherto sacred law such as the second law of thermodynamics became subject to possible violation—though with a very low probability, to be sure. Nevertheless, the mere possibility was shocking. As the twentieth century dawned, the edifice based on strict causality was pulled down by two revolutionaries, Einstein and Bohr, its place taken by the probabilistic architecture of quantum mechanics designed by Heisenberg, Schrödinger, and Dirac, leading to a wrenching disorientation of a large part of entrenched philosophy.

But another feature of quantum mechanics turned out to play a much more important role in the development of physics over the course of the second half of the twentieth century, a feature introduced into the theory right from the beginning by Bohr's model of the atom: its ability to sharply characterize a physical system by means of discrete energy levels—according to Einstein's special theory of relativity this implies masses—and other quantum numbers. If the primary aim of physicists before the second revolution was to describe and understand the motions, the dynamics, of physical entities—an aim at which they succeeded admirably—their principal aim after this paradigm shift changed to understanding the architecture of physical systems rather than their behavior. The goal of physics shifted from explaining *change* to explaining *being.*

Whereas classical physics, including relativity, had been, and still is, extremely reliable at predicting the future motion of objects such as planets and other heavenly bodies, as well as of rockets and space probes, it had never been very successful at elucidating structures. For instance, it had never succeeded in explaining the structure of the solar system, that is, why the distances between the various planets and the sun are what they are. From the perspective of classical physics, such an explanation fell outside its purview. Similarly, all attempts to explain, on the basis of classical electromagnetic theory, why the mass of the electron had the value experiments revealed did not succeed. Even the existence of atoms, half-heartedly accepted by some classical physicists, either enthusiastically defended or heatedly denied by others, was not fully demonstrated until the new physics came into being. Classical physics could, in any event, do no more than acknowledge atoms, while the new physics was able to explain why they existed, why they were stable, and why the atoms of a given element were all alike.

Quantum mechanics, of course, is also basically a dynamical theory, its equations able to predict the future course of the state of a physical system (not of individual events, though, as the state of a system is defined probabilistically), and the equations of quantum field theories such as those of the standard model of elementary particles define the dynamics of the fields. However, to the extent that these equations are approximately solvable, as in QED, they are rarely used for the purpose of predicting motions (exceptions being calculations of statistical predictions in the form of scattering and production cross sections). The equations are used predominantly, and in the case of QED extremely successfully, for predicting *structure*—atomic and nuclear energy levels, their multiplicities, and the resulting electromagnetic spectra, including the precise colors of the light emitted by the stars. Since motion and events can no longer be reliably predicted, the primary focus of physics now falls on what any quantum-mechanically based theory excels at, namely, explaining why there are families of particles with certain specific masses and other qualities, and why solid or fluid matter has the special properties it has. (Cosmology, which includes the description of large-scale motions, is an exception to this general statement, but here the underlying schema is primarily the classical theory of relativity rather than quantum mechanics.)

With a de-emphasis on dynamics in the new physics and a strong focus on why certain entities exist, it was inevitable that underlying abstract mathematical principles would play a dominant role, and symmetries as fundamental agents served that purpose very well. The cause of acceleration in Newton's mechanics was an anthropomorphically pushing force; the cause of the existence and anatomy of quarks, baryons, mesons, and leptons was a set of invariance principles. In the language of ancient Greek philosophy, there was a pronounced change from searching for Aristotelian efficient causes acting on given objects to seeking Platonic formal causes for the existence of these objects.

This shift is clearly visible from the heyday of QED to the present search for a "final theory." The very basis of the renormalization program that turned QED into the highly successful theory it became had been the trick of inserting "by hand" the particle masses (and some other parameters) rather than expecting the theory to predict them. In striking contrast, the theorists searching for the "theory of everything" are seeking the holy grail of a theory uniquely defined by an underlying general mathematical principle—a symmetry—able to predict the masses and other properties of all the fundamental constituents of the universe. If the grail is ever found, Aristotle's goal of understanding the relations between facts found by patient observation and experimentation will have lost out to Plato's approach to nature, which prized eternal intellectual, abstract principles over raw empiricism.

Does the history of physics outlined in this volume indicate that science "makes progress" in the sense of ultimately arriving at the truth about nature? Or can the activities of scientists over the last six thousand years be described as a mere succession of paradigm shifts, with no discernible direction, as the science historian Thomas Kuhn suggested in his influential book *The Structure of Scientific Revolution*? Readers will have to answer this question for themselves. My own view is that while physics has not arrived at *the truth* about the universe, nor will it ever, our activities are surely not aimless meanderings from one paradigm to another. Science has progressed, in the sense of ever more closely approaching an understanding of the workings of nature, even though the mathematics we use to describe this understanding may not turn out to be uniquely determined by nature itself. The twenty-first century will no doubt get us even closer to this kind of truth, either by way of Aristotle or Plato.

Notes

1. Beginnings

1. Sarton, *A History of Science: Ancient Science through the Golden Age of Greece,* pp. 35–36.

2. The Greek Miracle

1. Needham, *Science and Civilisation in China,* vol. 2, p. 55.
2. Ibid., p. 63.
3. Sarton states that Hindu philosophers engaged in similar discussions, but their dates seem to be extremely uncertain, and no influence either way need be assumed. Sarton, *A History of Science: Hellenistic Science and Culture in the Last Three Centuries* B.C., p. 248.
4. Sarton regards the evidence for this as unconvincing, but his writings indicate that he was somewhat prejudiced against Phoenicians.
5. Lloyd and Sivin, *The Way and the Word,* p. 190.
6. Needham, vol. 2, p. 124.

3. Science in the Middle Ages

1. Grant, *The Foundations of Modern Science in the Middle Ages.*
2. Rubenstein, *Aristotle's Children,* p. 86.
3. Grant, p. 21.
4. Huff, *The Rise of Early Modern Science,* pp. 149ff.
5. Translated in David Knowles, *The Evolution of Medieval Thought,* p. 200, quoted by Grant, pp. 68–69.
6. Aristotle, *On the Heavens* 2.1.283b.26–30, translated by W. K. C. Guthrie, quoted by Grant, p. 74.
7. Grant, p. 75.

8. Thomas Aquinas, *Summa Theologiae,* 1.46.1, in *St. Thomas Aquinas, Siger of Brabant, St. Bonaventure, On the Eternity of the World (De aeternitate mundi),* 66; quoted by Grant, p. 76.
9. See Grant, pp. 76f.
10. Sarton, p. 626.
11. Sarton, p. 507.
12. See Huff, p. 47, n. 2, for one possible explanation; see also pp. 211ff.
13. See Grant, *Nicole Oresme and the Kinematics of Circular Motion,* and R. Small, "Incommensurability and recurrence."
14. Jervis, *Cometary Theory in Fifteenth-Century Europe,* p. 12.
15. Jervis, pp. 91ff.

4. The First Revolution

1. Holton, *Victory and Vexation in Science,* p. 164.
2. R. Newton, *What Makes Nature Tick?* p. 212.
3. R. Newton, *Galileo's Pendulum: From the Rhythm of Time to the Making of Matter.*
4. For the convoluted history of this law, see Brush, *The Kind of Motion We Call Heat,* p. 12.
5. Westfall, *The Life of Isaac Newton,* p. 14.
6. Ibid., p. 40.
7. Quoted in Newman, *The World of Mathematics,* p. 277.
8. Brush, p. 14.
9. Merton, *On the Shoulders of Giants,* p. 1.
10. Aiton, *Leibniz: A Biography,* pp. 341ff.

5. Newton's Legacy

1. Laplace, *Essai sur les probabilités,* p. 4.
2. See Diacu and Holmes, *Celestial Encounters.*

7. Relativity

1. Pais, *Subtle Is the Lord,* pp. 211–212.
2. H. Kragh and R. W. Smith, "Who discovered the expanding universe?" *History of Science* 41 (2003): 141–162.
3. Einstein, "Kosmologische Betrachtungen zur allgemeined Relativitätstheorie," *Sitzungsberichte der Kgl. Preussische Akademie der Wissenschaften,* 1917, pp. 142–152.

8. Statistical Physics

1. J. W. Gibbs, *Proceedings of the American Academy of Arts & Sciences* 16

(1889): 458, and *Scientific Papers,* vol. 2, p. 265.

2. Gibbs, *Elementary Principles in Statistical Mechanics.*
3. For a translation, see *Physics Today* 45, no. 1 (January 1992): 44.
4. For a demonstration of this, see R. Newton, *Thinking about Physics,* pp. 139ff.

9. Probability

1. Hacking, Ian. *The Emergence of Probability,* p. 52.
2. Ibid., p. 112.
3. Quoted on p. 245 of von Plato, *Creating Modern Probability,* from de Finetti, *Bolletino dell'Unione Matematica Italiana,* No. 5 (1930), pp. 258–261.

10. The Quantum Revolution

1. See Smith, "J. J. Thomson and the electron," in Buchwald and Warwick, *Histories of the Electron: The Birth of Microphysics,* pp. 21–76.
2. Chadwick, *Nature* 140 (1937): 749.
3. Furthermore, in 1922, the two German physicists Otto Stern and Walther Gerlach had discovered that when an electron, which they regarded as a tiny magnet, was sent through an inhomogeneous magnetic field so that the magnetic force would deflect it depending on the direction of its magnetic momentum, the angular deflections were quantized: they occurred only in specific, discrete directions. No matter how they oriented their apparatus with respect to a given incident beam, the electrons emerged in two separate directions with respect to it, one ray bent upward and one down. After the discovery of the electron's spin, which had to be aligned with its magnetic moment, it was clear that what Stern and Gerlach had discovered was the quantization of angular momentum and the fact that the magnitude of the electron's spin was ½ of Planck's constant. This was the physical origin of Pauli's two-valued fourth atomic quantum number.
4. Schweber, *QED and the Men Who Made It,* p. 70.
5. Quoted by Petersen, "The philosophy of Niels Bohr," *Bulletin of the Atomic Scientists,* September 1963, p. 12.
6. Ibid., p. 12.
7. Kemble, *The Fundamental Principles of Quantum Mechanics,* p. 33. *Scientific American,* November 1979, p. 158.
8. See Omnes, *Understanding Quantum Mechanics.*
9. Aspect et al., *Physical Review Letters* 49 (1982): 1804.

Sources and Further Reading

Achinstein, P. *The Nature of Explanation.* New York: Oxford University Press, 1983.

Aczel, Amir D. *Entanglement: The Greatest Mystery in Physics.* New York: Four Walls Eight Windows, 2001.

Aiton, E. J. *Leibniz: A Biography.* Boston: Hilger, 1985.

Andrade, E. N. da C. *Rutherford and the Nature of the Atom.* Garden City, NY: Doubleday & Co., Inc., 1964.

Bernal, J. D. *Science in History.* New York: Hawthorn Books, Inc., 1965.

Bhōskarōcōrya. *Līlōvatī of Bhōskarōcōrya.* Delhi: Motilal Banarsidass Publishers, 2001.

Born, Max. *Natural Philosophy of Cause and Chance.* New York: Dover, 1964.

———. *Physics in My Generation.* New York: Springer Verlag, 1969.

Boyer, Carl B. *A History of Mathematics.* New York: John Wiley & Sons, Inc., 1968.

Brown, L. M., A. Pais, and B. Pippard. *Twentieth Century Physics.* Bristol: IOP Publishing Ltd., 1995.

Brush, Stephen G. *The Kind of Motion We Call Heat: A History of the Kinetic Theory of Gases in the 19th Century,* Book 1. Amsterdam: North Holland Publishing Co., 1976.

Buchwald, Jed Z. and Andrew Warwick. *Histories of the Electron: The Birth of Microphysics.* Cambridge, MA: MIT Press, 2001.

Bunge, Mario, ed. *Quantum Theory and Reality.* New York: Springer Verlag, 1967.

Burchfield, Joe D. *Lord Kelvin and the Age of the Earth.* New York: Science History Publications, 1975.

Calinger, Ronald, ed. *Classics of Mathematics.* Oak Park, IL: Moore Publishing Co., 1982.

Cercignani, Carlo. *Ludwig Boltzmann: The Man Who Trusted Atoms.* New York: Oxford University Press, 1998.

Cohen, I. Bernard. *The Birth of a New Physics,* revised and updated ed. New York: W. W. Norton & Co., 1985.

Cook, David B. *Probability and Schrödinger's Mechanics.* Singapore: World Scientific Publishing Co., 2002.

Cropper, William H. *Great Physicists: The Life and Times of Leading Physicists from Galileo to Hawking.* New York: Oxford University Press, 2001.

David, F. N. "Some notes on Laplace" in J. Neyman and L. M. LeCam, eds. *Bernoulli, Bayes, Laplace.* New York: Springer Verlag, 1965, pp. 30–44.

de Finetti, Bruno. *Theory of Probability,* 2 vols. Chichester: John Wiley & Sons, 1990.

Diacu, Florin and Philip Holmes. *Celestial Encounters: The Origins of Chaos and Stability.* Princeton, NJ: Princeton University Press, 1996.

Dirac, P. A. M. *The Principles of Quantum Mechanics,* 3rd ed. Oxford: Clarendon Press, 1947.

Dyson, Freeman. "The death of a star." *Nature,* 29 December 2005, p. 1086.

Ehrenberg, W. *Dice of the Gods: Causality, Necessity, and Chance.* London: Birkbeck College, 1977.

Fermi, Laura. *Atoms in the Family.* London: Allen and Unwin, 1955.

Frank, P. et al., eds. *Selected Papers of Richard von Mises,* vol. 1. Providence, RI: American Mathematical Society, 1963.

Franklin, Allan. *Experiment, Right or Wrong.* Cambridge: Cambridge University Press, 1990.

Galison, Peter. *How Experiments End.* Chicago: University of Chicago Press, 1987.

Glashow, Sheldon. *Interactions: A Journey Through the Mind of a Particle Physicist and the Matter of This World.* New York: Warner Books, 1988.

Grandy, R. E., ed. *Theories and Observation in Science.* Englewood Cliffs, NJ: Prentice-Hall, 1973.

Grant, Edward. *The Foundations of Modern Science in the Middle Ages: Their Religious, Institutional, and Intellectual Contexts.* Cambridge: Cambridge University Press, 1996.

———. *Nicole Oresme and the Kinematics of Circular Motion.* Madison, WI: University of Wisconsin Press, 1971.

Greene, Brian. *The Fabric of the Cosmos: Space, Time, and the Texture of Reality.* New York: Knopf, 2004.

Greenspan, Nancy Thorndike. *The End of the Certain World: The Life and Science of Max Born.* New York: Basic Books, 2005.

Hacking, Ian. *The Emergence of Probability: A Philosophical Study of Early Ideas about Probability, Induction, and Statistical Inference.* Cambridge: Cambridge University Press, 1975.

Heilbron, J. L. *The Dilemmas of an Upright Man: Max Planck as Spokesman of German Science.* Berkeley, CA: University of California Press, 1986.

———. *Electricity in the 17th and 18th Centuries: A Study in Early Modern Physics.* Mineola, NY: Dover Publications, 1999.

Heisenberg, Werner. *The Physical Principles of the Quantum Theory.* Chicago: University of Chicago Press, 1930.

Holton, Gerald. *Victory and Vexation in Science: Einstein, Bohr, Heisenberg, and Others.* Cambridge, MA: Harvard University Press, 2005.

Hon, G. and S. S. Rakover, eds. *Explanation: Theoretical Approaches and Applications.* Dordrecht, Netherlands: Kluwer Academic Publishing, 2001.

Huff, Toby E. *The Rise of Early Modern Science: Islam, China, and the West,* 2nd ed. Cambridge: Cambridge University Press, 2003.

Ihsanoglu, Ekmeleddin. *Science, Technology, and Learning in the Ottoman Empire: Western Influence, Local Institutions, and the Transfer of Knowledge.* Burlington, VT: Ashgate, 2004.

Jammer, Max. *The Philosophy of Quantum Mechanics.* New York: John Wiley & Sons, 1974.

Jauch, Joseph. *Foundations of Quantum Mechanics.* Reading, MA: Addison Wesley, 1968.

Jervis, Jane L. *Cometary Theory in Fifteenth-Century Europe.* Wroclaw, Poland: The Polish Academy of Sciences Press, 1985.

Kaempffer, F. A. *Concepts in Quantum Mechanics.* New York: Academic Press, 1965.

Kaye, G. R. *Indian Mathematics.* Calcutta: Thacker, Spink & Co, 1915.

Keil, F. C. and R. A. Wilson, eds. *Explanation and Cognition.* Cambridge, MA: MIT Press, 2000.

Kemble, E. C. *The Fundamental Principles of Quantum Mechanics.* New York: McGraw-Hill, 1937.

Kolmogorov in Perspective. Providence, RI: American Mathematical Society, 2000.

Kuhn, Thomas S. *The Structure of Scientific Revolution*, 2nd ed. Chicago: University of Chicago Press, 1970.

Livi, R. and A. Vulpiani. *The Kolmogorov Legacy in Physics.* Berlin: Springer Verlag, 2003.

Lloyd, G. and N. Sivin. *The Way and the Word: Science and Medicine in Early China and Greece.* New Haven, CT: Yale University Press, 2002.

Merton, Robert K. *On the Shoulders of Giants.* Chicago: University of Chicago Press, 1993.

Needham, J. *Science and Civilisation in China*, vol. 2. Cambridge: Cambridge University Press, 1956.

———. *Science and Civilisation in China*, vol. 4, Part I. Cambridge: Cambridge University Press, 1962.

Newman, James R. *The World of Mathematics.* New York: Simon and Schuster, 1956.

Newton, R. G. *Galileo's Pendulum: From the Rhythm of Time to the Making of Matter.* Cambridge: Harvard University Press, 2004.

———. *Thinking about Physics.* Princeton, NJ: Princeton University Press, 2000.

———. *What Makes Nature Tick?* Cambridge: Harvard University Press, 1998.

Omnes, Roland. *Understanding Quantum Mechanics.* Princeton, NJ: Princeton University Press, 1994.

Pais, A. *Niels Bohr's Times: In Physics, Philosophy, and Polity.* Oxford: Clarendon Press, 1991.

———. *Subtle Is the Lord . . . : The Science and the Life of Albert Einstein.* New York: Oxford University Press, 1982.

Parameswaran, S. *The Golden Age of Indian Mathematics.* Kerala, India: Swadeshi Science Movement, 1998.

Pitt, J. C. *Theories of Explanation.* New York: Oxford University Press, 1988.

Popper, K. R. *The Logic of Scientific Discovery.* New York: Basic Books, 1959.

———. *Realism and the Aim of Science.* Totowa, NJ: Rowman and Littlefield, 1983.

Press, S. J. and J. M. Tanur. *The Subjectivity of Scientists and the Bayesian Approach.* New York: John Wiley & Sons, 2001.

Rao, S. Balachandra. *Aryabhata-I and His Astronomy.* Tirupati, India: Rashtriya Sanskrit Vidyapeetha, 2003.

———. *Bhaskara-I and His Astronomy.* Tirupati, India: Rashtriya Sanskrit Vidyapeetha, 2003.

———. *Indian Astronomy: An Introduction.* Hyderabad, India: Universities Press, 2000.

———. *Indian Mathematics and Astronomy: Some Landmarks,* 2nd ed. Bangalore, India: Jnana Deep Publications, 1998.

Reichenbach, Hans. *Philosophic Foundations of Quantum Mechanics.* Berkeley, CA: University of California Press, 1946.

Robinson, Andrew. *The Last Man Who Knew Everything: Thomas Young.* New Jersey: Pi Press, 2006.

Rosenkrantz, R. D., ed. *E. T. Jaynes: Papers of Probability, Statistics, and Statistical Physics.* Dordrecht, Netherlands: D. Reidel Publishing Co., 1983.

Roy, S. B. *Prehistoric Lunar Astronomy.* New Delhi: Institute of Chronology, 1976.

Ruben, D. H., ed. *Explanation.* New York: Oxford University Press, 1993.

Rubenstein, Richard E. *Aristotle's Children: How Christians, Muslims, and Jews Rediscovered Ancient Wisdom and Illuminated the Dark Ages.* New York: Harcourt, Inc., 2003.

Ruhla, Charles. *The Physics of Chance.* Oxford: Oxford University Press, 1989.

Salmon, W. C. *Causality and Explanation.* New York: Oxford University Press, 1998.

———. *Scientific Explanation and the Causal Structure of the World.* Princeton, NJ: Princeton University Press, 1984.

Sarkar, Benoy Kumar. *Hindu Achievements in Exact Science: A Study in the History of Scientific Development.* New York: Longmans, Green and Co., 1918.

Sarton, George. *A History of Science: Ancient Science through the Golden Age of Greece.* Cambridge, MA: Harvard University Press, 1952.

———. *A History of Science: Hellenistic Science and Culture in the Last Three Centuries B.C.* Cambridge, MA: Harvard University Press, 1959.

———. *Introduction to the History of Science, Vol. I, From Homer to Omar Khayyam.* Baltimore, MD: The Williams & Wilkins Co., 1927.

———. *Introduction to the History of Science, Vol. II, Part I, From Rabbi Ben Esra to Ibn Rushd.* Baltimore, MD: The Williams & Wilkins Co., 1931.

———. *Introduction to the History of Science, Vol. II, Part II, From Robert Grosseteste to Roger Bacon.* Baltimore, MD: The Williams & Wilkins Co., 1931.

———. *Introduction to the History of Science, Vol. III, Science and Learning in the Fourteenth Century, Part I, First Half of the Fourteenth Century.* Baltimore, MD: The Williams & Wilkins Co., 1947.

———. *Introduction to the History of Science, Vol. III, Science and Learning in the Fourteenth Century, Part II, Second Half of the Fourteenth Century.* Baltimore, MD: The Williams & Wilkins Co., 1948.

Schweber, Silvan S. *QED and the Men Who Made It: Dyson, Feynman, Schwinger, and Tomonaga.* Princeton, NJ: Princeton University Press, 1994.

Selin, Helaine. *Astronomy Across Cultures: The History of Non-Western Astronomy.* Dordrecht, Netherlands: Kluwer Academic Publishing, 2000.

Small, Robin. "Incommensurability and recurrence: from Oresme to Simmel." *Journal of the History of Ideas,* 52, No. 1, pp. 121–137, 1991.

Somayaji, D. A. *A Critical Study of the Ancient Hindu Astronomy in the Light and Language of the Modern.* Dharwar, India: Karnatak University, 1971.

Suppes, Patrick. *A Probabilistic Theory of Causality.* Amsterdam: North Holland Publishing Co., 1970.

Suter, Heinrich. *Die Mathematiker und Astronomen der Araber und ihre Werke.* Leipzig, Germany: B. G. Teubner, 1900.

Swinburne, R. *Simplicity as Evidence of Truth.* Milwaukee: Marquette University Press, 1997.

Thorndike, Lynn. *Science and Thought in the Fifteenth Century.* New York: Columbia University Press, 1929.

von Plato, Jan. *Creating Modern Probability: Its Mathematics, Physics and Philosophy in Historical Perspective.* Cambridge: Cambridge University Press, 1994.

Weinberg, Steven. *Dreams of a Final Theory: The Search for the Fundamental Laws of Nature.* New York: Pantheon Books, 1992.

Westfall, Richard S. *The Construction of Modern Science: Mechanisms and Mechanics.* Cambridge: Cambridge University Press, 1977.

———. *The Life of Isaac Newton.* Cambridge: Cambridge University Press, 1993.

———. *Never at Rest: A Biography of Isaac Newton.* Cambridge: Cambridge University Press, 1980.

Wilson, Robert. *Astronomy Through the Ages, The Story of the Human Attempt to Understand the Universe.* Princeton, NJ: Princeton University Press, 1997.

Witten, Edward. "Unraveling string theory." *Nature,* 29 December 2005, p. 1085.

Wright, J. *Realism and Explanatory Priority.* Dordrecht, Netherlands: Kluwer Academic Publishing, 1997.

Index